Coniacian and Santonian belemnite faunas from Bornholm, Denmark

WALTER KEGEL CHRISTENSEN AND MAX-GOTTHARD SCHULZ

Christensen, W.K. & Schulz, M.-G. 1997 08 22: Coniacian and Santonian belemnite faunas from Bornholm, Denmark. *Fossils and Strata*, No. 44, pp. 1–73. Oslo. ISSN 0300-9491. ISBN 82-00-37695-8.

The belemnite faunas of the Lower–Middle Coniacian Arnager Limestone and Upper Coniacian – Lower Santonian Bavnodde Greensand Formations of the island of Bornholm, Denmark, are described, using biometric analysis. The fauna of the Arnager Limestone Formation includes four taxa: *Actinocamax verus* cf. *antefragilis* Naidin, new for Bornholm, *Goniocamax lundgreni lundgreni* (Stolley), *Goniocamax* sp.nov.(?) aff. *lundgreni*, and *Goniocamax(?)* sp.nov.(?). The fauna of the Bavnodde Greensand Formation is remarkably diverse and comprises nine taxa: *Actinocamax verus verus* Miller, *Gonioteuthis praewestfalica* Ernst & Schulz, new for Bornholm, *Gonioteuthis westfalica westfalica* (Schlüter), *Gonioteuthis ernsti* sp.nov., *Goniocamax lundgreni lundgreni* (Stolley), *Goniocamax birkelundae* sp.nov., *Goniocamax striatus* sp.nov., *Belemnitella schmidi* sp.nov., and *Belemnitella propinqua propinqua* (Moberg). Three belemnite zones and eight belemnite assemblage zones are recognized. The Coniacian has a Zone of *Goniocamax lundgreni* and includes three assemblage zones. The Lower Santonian has a Zone of *Gonioteuthis praewestfalica* below and a Zone of *Gonioteuthis westfalica* above. The lower zone is subdivided into two and the upper zone into three assemblage zones. The belemnite faunas include species from the Central European and Central Russian Subprovinces of the North European palaeobiogeographical Province. A major faunal turnover of the belemnite faunas takes place at the base of the *G. westfalica* Zone. *Goniocamax lundgreni* predominates below and *Gonioteuthis westfalica* above. □Belemnitellids, ACTINOCAMAX, GONIOCAMAX, GONIOTEUTHIS, BELEMNITELLA, Upper Cretaceous, Coniacian, Santonian, taxonomy, biometry, biostratigraphy, palaeobiogeography, Arnager Limestone Formation, Bavnodde Greensand Formation, Bornholm, Denmark.

Walter K. Christensen [wkc@savik.geomus.ku.dk], Geological Museum, University of Copenhagen, Øster Voldgade 5–7, DK-1350 Copenhagen, Denmark; Max-Gotthard Schulz, Geologisch-Paläontologisches Institut der Universität, Olshausenstraße 40, D-24118 Kiel, Germany; 9th June, 1995; revised 8th February, 1996.

Contents

Introduction

The belemnites of the Upper Cretaceous Arnager Greensand, Arnager Limestone, and Bavnodde Greensand Formations of the island of Bornholm, Denmark, were monographed by Birkelund (1957). In addition to this paper belemnites from Bornholm were described, recorded, and/or used biostratigraphically by Schlüter (1874, 1876), Stolley (1897, 1930), Ravn (1916, 1918, 1921, 1930, 1946), Jeletzky (1958), and more recently by Christensen (1971, 1973, 1985, 1990a, 1991), Kennedy & Christensen (1991), and Tröger & Christensen (1991).

The Lower–Middle Coniacian Arnager Limestone and Upper Coniacian – Lower Santonian Bavnodde Greensand Formations have yielded ammonites (Kennedy & Christensen 1991), inoceramid bivalves (Tröger & Christensen 1991), foraminifera (Douglas & Rankin 1969; Bailey & Hart 1979; Solakius & Larsson 1985; Solakius 1988, 1989; Packer et al. 1989), and dinoflagellate cysts (Packer et al. 1989; Schiøler 1992), in addition to belemnites.

During the Coniacian – Lower Campanian, two belemnite subprovinces within the North European palaeobiogeographical province have been recognized. The Central European Subprovince is characterized by the *Gonioteuthis* lineage and the Central Russian Subprovince by the *Goniocamax–Belemnitella* lineage (Christensen 1975a, 1976, 1988, 1990b, 1991). The Baltoscandian belemnite faunas from this time interval are of great significance, because they include species from both subprovinces, thus providing a basis for correlation.

The aim of the present paper is to describe the Coniacian and Santonian belemnite faunas of Bornholm, utilizing biometric methods, and to place the Arnager Limestone and Bavnodde Grensand Formations as well as the outcrops in the international stratigraphical framework.

Material

The majority of the belemnites examined during the course of the present study were collected in the field by the authors aided by German and Danish colleagues. These belemnites are very accurately horizoned with reference to lithological marker beds. The material is housed in the Geological Museum, University of Copenhagen, and the Geologisch-Paläontologisches Institut der Universität Kiel. Collections made in the last part of the 19th century and the first half of this century, including the material described by Stolley (1897), Ravn (1918, 1921, 1946), and Birkelund (1957), which are housed in the Geological Museum, Copenhagen, are also included in the present study. These belemnites usually are labelled only with the locality name. Additional belemnites have been provided by Danish and German colleagues. We have also obtained specimens on loan from the following institutions: Geologisch-Paläontologisches Institut, Greifswald, Geological Survey of Sweden, Uppsala, Geological Institute, Lund, Swedish Museum of Natural History, Section of Paleozoology, Stockholm, and Geological Survey of Canada, Ottawa.

The following abbreviations are used to indicate the location of specimens mentioned in the text: GM and MGUH, Geological Museum, Copenhagen; GPIG, Geologisch-Paläontologisches Institut, Greifswald; GPIH, Geologisch-Paläontologisches Institut, Hamburg; GPIK, Geologisch-Paläontologisches Institut, Kiel; GSC, Geological Survey of Canada, Ottawa; LM, Geological Institute, Lund; RM, Naturhistoriska Riksmuseet, Stockholm (Swedish Museum of Natural History, Section of Paleozoology); SGU, Sveriges Geologiska Undersökning, Uppsala (Geological Survey of Sweden).

Geological setting

The island of Bornholm in the Baltic Sea is a horst situated within the Fennoscandian Border Zone, i.e. the marginal area between the stable Precambrian Baltic Shield and the subsiding Late Palaeozoic and Mesozoic Danish Subbasin. The northern part of the island consists of Precam-

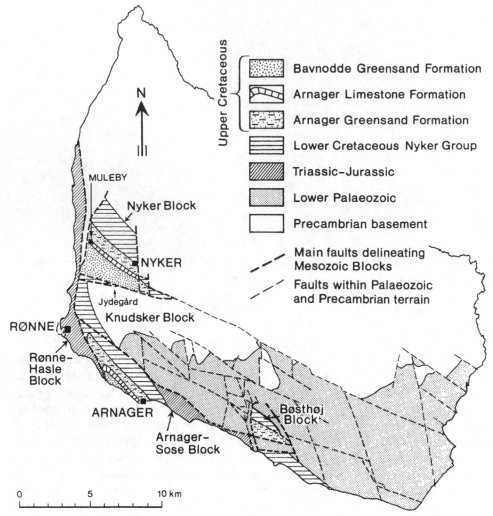

Fig. 1. Geological map of Bornholm. Modified from Gry (1960, 1969) and Gravesen *et al.* (1982).

brian basement, and Palaeozoic and Mesozoic sedimentary strata occur in down-faulted blocks to the south and west (Fig. 1).

The marine Cretaceous sediments, ranging in age from the Albian to the Santonian, include the following lithostratigraphical formations, bottom to top: the lower Middle Cenomanian Arnager Greensand *s.str.*, the Lower–Middle Coniacian Arnager Limestone, and the Upper Coniacian – Lower Santonian Bavnodde Greensand (Figs. 2–3). The phosphatic conglomerate near the base of the Arnager Greensand contains derived Lower Albian and Lower Cenomanian ammonites.

The formations occur in three fault blocks: the Nyker block to the west, the Arnager–Sose block to the southwest, and the Bøsthøj block to the south (Fig. 1). The Arnager Limestone and Bavnodde Greensand Formations are recorded only from the Nyker and Arnager–Sose blocks. Christensen *in* Tröger & Christensen (1991) reviewed the formations and provided locality details.

The Upper Cretaceous deposits of Bornholm are known from low coastal cliffs on the southwest coast, small sections along stream-cuttings, small inland pits, most of which are abandoned, and shallow bore-holes. The best exposures of the sedimentary succession are found now in the low coastal cliffs on the southwest coast, from Bavnodde in the west to Madsegrav in the east. The thickness of the succession is estimated to be about 200 m, and the succession consists mainly of detrital clastic sediments: calcareous, glauconitic, clayey quartz sand/sandstones and siltstones.

The strike of the base of the Arnager Limestone west of Arnager is 140°, and the dip is 8°SW. The strike of the top of the formation east of Horsemyre Odde is 120°, and the dip is 8°SW. The strike of the Bavnodde Greensand east of Bavnodde is 135°, and the dip is 10°SW. The strike is thus almost parallel to the main direction of the southwest coast.

Since the oldest beds of the Upper Cretaceous succession are exposed to the east at Madsegrav and the young-

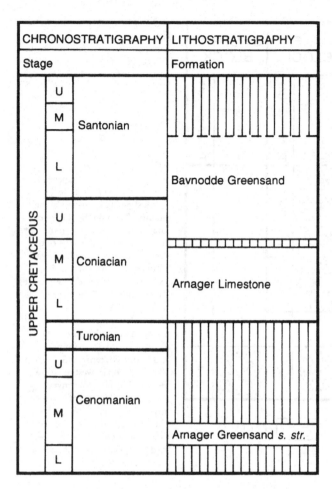

CHRONOSTRATIGRAPHY			LITHOSTRATIGRAPHY
Stage			Formation
UPPER CRETACEOUS	Santonian	U	
		M	
		L	Bavnodde Greensand
	Coniacian	U	
		M	Arnager Limestone
		L	
	Turonian		
	Cenomanian	U	
		M	
			Arnager Greensand *s. str.*
		L	

Fig. 2. Stratigraphical scheme of the Upper Cretaceous of Bornholm.

est beds crop out to the west at Bavnodde (Fig. 4), it was suggested by earlier authors, including Birkelund (1957, p. 29), that the Bavnodde Greensand exposed on the southwest coast becomes progressively younger westwards from Horsemyre Odde to Bavnodde. This suggestion is incorrect. First, because of the strike and dip of the sediments and the course of the coastline, the greensand becomes older from Horsemyre Odde westwards to east of Forchhammers Odde (see below). Second, the belemnite assemblage from Forchhammers Odde is younger than that from east of Forchhammers Odde (see below), implying that there is a fault immediately east of Forchhammers Odde (Fig. 4). We also suggest that there is a fault east of Horsemyre Odde (Fig. 4), because the base of the Arnager Greensand is ca. 26 m deeper in the two western boreholes (DGU Nos 246.615 and 246.617) than in the eastern borehole (DGU No. 246.616).

Because of these faults, there are two stratigraphical gaps in the Bavnodde Greensand succession exposed on the southwest coast: A minor gap near the base of the greensand and a larger gap, corresponding to the lower part of the Lower Santonian, i.e. between the lower part of the greensand, exposed east of Forchhammers Odde and eastwards, and the upper part of the greensand, exposed at Forchhammers Odde and westwards. However, lower Lower Santonian greensand is exposed elsewhere.

Arnager Greensand Formation

The formation is about 85 m thick (Tröger & Christensen 1991). The Arnager Greensand *sensu stricto* has yielded the belemnites *Actinocamax primus* Arkhangelsky and *A.* aff./cf. *primus* (Christensen 1990a), ammonites (Kennedy, Hancock & Christensen 1981), inoceramid bivalves (Tröger & Christensen 1991), in addition to foraminifera and dinoflagellate cysts (Packer *et al.* 1989) of early Middle Cenomanian age. Derived ammonites from the phosphatic conglomerate near the base of the formation are from the Lower Albian *Leymeriella tardefurcata* and *Douvilleiceras mammillatum* Zones, and the Lower Cenomanian *Mantelliceras saxbii* and/or *M. dixoni* Zones (Kennedy *et al.* 1981). This formation will not be considered further, because the age is well-known and the belemnites were described elsewhere.

Arnager Limestone Formation

This formation is sandwiched between the Arnager Greensand and Bavnodde Greensand. The formation has generally a higher content of carbonate than the subjacent and superjacent formations. It is, however, only on the southwest coast immediately west of Arnager that the formation is developed as siliceous, marly limestone. At this site, deposition of the limestone took place in a complex of low mud-mounds most likely caused by the baffling effect of siliceous sponges working in concert with sponge spicule mats (Noe-Nygaard & Surlyk 1985). According to Hamann (1988, Fig. 32G), the limestone is a lens-shaped body of rock enclosed in detrital clastic sediments. This lens-shaped body developed along the crest of the Arnager–Sose fault block, which may have formed a structural high during the Lower Coniacian. Gry (1960) estimated the thickness of the formation on the southwest coast to be 12–20 m, of which 7–8 m are exposed.

The formation consists of calcareous siltstones, sandy limestone, glauconitic marl, or sandy, glauconitic marl northwest and southeast of Arnager. According to the geological map of Hamann (1987; see also Hamann 1989) the thickness of the formation increases northwest of Horsemyre Odde. On the basis of a presumed general dip of 8° to the southwest, we estimate the thickness to be 35–40 m in the Stampe Å area. This suggestion is consistent with the biostratigraphical results, because the formation is lower Lower Coniacian from Arnager to east of Horse-

| Lithostratigraphy | ARNAGER–SOSE BLOCK | | | | NYKER BLOCK | | BØSTHØJ BLOCK | |
| | Southwest coast | | Stampe Å | | | | | |
	Lithology	Thickness (m)	Lithology	Thickness (m)	Lithology	Thickness (m)	Lithology	Thickness (m)
Bavnodde Greensand Formation		c. 70				?		
Arnager Limestone Formation		12-20		35-40		?		
Arnager Greensand Formation		85		?		?		?

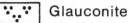 Sandstone   Conglomerate   Marl

Limestone   Glauconite

Fig. 3. Scheme showing the Cenomanian–Santonian sedimentary sequences on three fault blocks on Bornholm. Modified from Tröger & Christensen (1991).

Fig. 4 (below on facing pages). Map of the Arnager–Sose fault block showing exposures along Stampe Å and on the southwest coast, from Madsegrav in the east to west of Bavnodde. The outcrops showing Bavnodde Greensand are abbreviated as follows: WB = west of Bavnodde; EB = east of Bavnodde; WF = west of Forchhammers Odde; F = Forchhammers Odde; EF = east of Forchhammers Odde; FS = between Forchhammers Odde and Skidteper; WS = west of Skidteper; WH = west of Horsemyre Odde; H = Horsemyre Odde; and EH = east of Horsemyre Odde. In the Stampe Å area, the Arnager Limestone Formation is exposed at loc. 4 and 5. Modified from Kennedy & Christensen (1991).

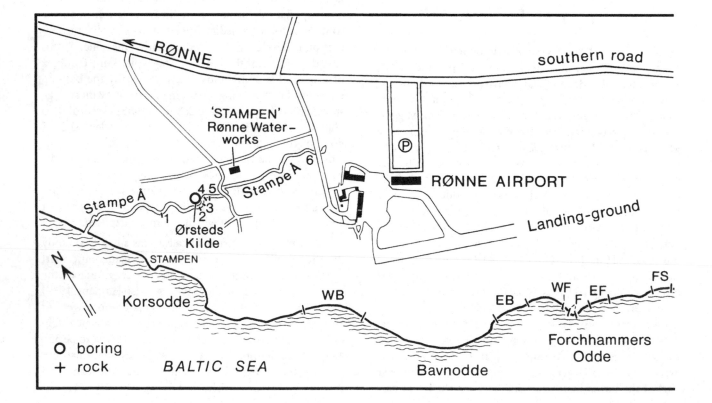

myre Odde, whereas the upper part of the formation at loc. 5 at Stampe Å is Middle Coniacian (see below).

Bavnodde Greensand Formation

An aggregate of 45–50 m of glauconitic, calcareous, fine-grained, silty quartz sands, with several layers of nodular greensand, quartzitic beds, and marly beds, are exposed in the low coastal cliffs on the southwest coast. On the basis of the geological map by Hamann (1987) and a presumed general dip of 8° to the southwest, we estimate the thickness of the formation, from its base to the outcrop west of Bavnodde, to be ca. 70 m. Gry (1960) and Brüsch (1984) estimated the thickness to be 180 m and 120 m, respectively.

Locality details

Arnager–Sose block

Southwest coast. – The lower 7 m of the Arnager Limestone are exposed immediately west of Arnager at its type locality and the top part of the formation, less than 1 m, and the basal part of the superjacent Bavnodde Greensand

is intermittently exposed ca. 155 m east of the point of Horsemyre Odde after winter storms have removed the shingle (Fig. 4). We have examined belemnites from the lower 7 m and the top part of the Arnager Limestone.

The Bavnodde Greensand crops out in scattered exposures in the low coastal cliffs from Horsemyre Odde to west of Bavnodde (Fig. 4). Sections of most of these outcrops are shown in Figs. 5–6. The greensand east of Horsemyre Odde crops out in very small scattered exposures along the foot of the cliff and on the shore platform.

Ravn (1921) introduced the locality name Forchhammers Klint, which includes the district from east of Bavnodde to Forchhammers Odde (Fig. 4). At that time the greensand was exposed along the entire cliff. Owing to later coast protection, the greensand is exposed now only at a few sites. Forchhammers Klint *sensu* Ravn thus includes the exposures at east of Bavnodde, west of Forchhammers Odde, and Forchhammers Odde as used herein.

Stampe Å area. – The Upper Cretaceous deposits in this area are poorly known, because of the small number of stream cuttings along the brook Stampe Å. A sandy limestone of the Arnager Limestone Formation is exposed at loc. 4, which has not yielded diagnostic fossils. Belemnites occur in the glauconitic marl of the Arnager Limestone Formation at loc. 5 (Fig. 4).

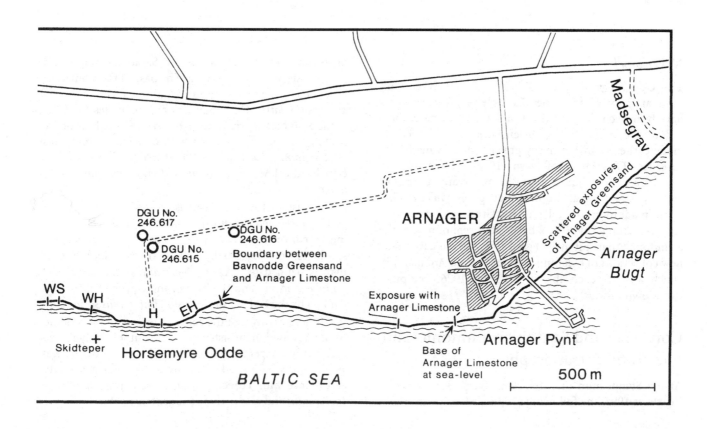

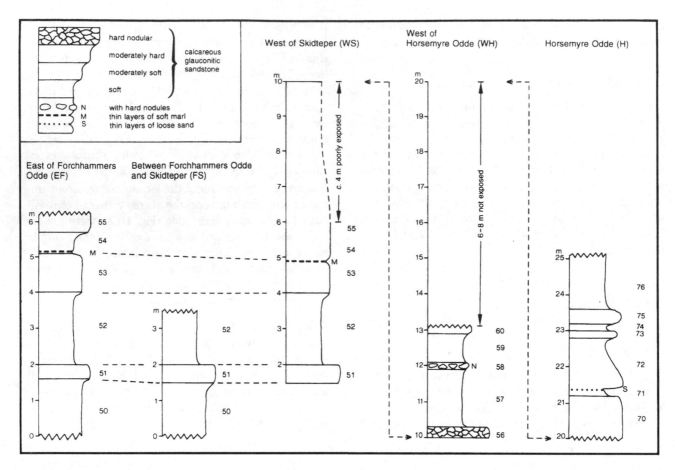

Fig. 5. Five sections of the Bavnodde Greensand from east of Forchhammers Odde to Horsemyre Odde, showing approximately the lower 25 m of the greensand.

Nyker block

The Upper Cretaceous sediments of the Nyker fault block are poorly known, because only very few exposures have been recorded and most of these are not accessible today. We have studied belemnites from collections made earlier from the marly greensands and marls, the so-called 'Glass-marl' of Ravn (1946), of the Arnager Limestone Formation exposed along stream cuttings of Muleby Å and the abandoned marl pit at Muleby (Figs. 7–8), in addition to the Bavnodde Greensand exposed at stream cuttings of Blykobbe Å at Risenholm and the abandoned marl pit at Jydegård (Fig. 9). We have made new collections of belemnites from the Jydegård marl pit, which was still accessible in 1975 and 1976. The pit is overgrown now.

Coniacian and Santonian ammonite and inoceramid stratigraphy

At the Symposium on Cretaceous Stage Boundaries in Bruxelles 1995, the following proposals for the definition

of the bases of the Coniacian and Santonian stages and their substages were suggested. The base of the Coniacian is defined by the first appearance datum (FAD) of *Inoceramus rotundatus*. This boundary approximates the FAD of the ammonite *Forresteria petrocoriensis*. The base of the Middle Coniacian is defined by the FAD of *Inoceramus (Volviceramus) koeneni*, and the base of the Upper Coniacian by the FAD of *Inoceramus (Magadiceramus) subquadratus*.

The base of the Santonian is defined by the FAD of *Inoceramus (Cladoceramus) undulatoplicatus*. The following provisional working definitions for the Santonian substages were suggested: The base of the Middle Santonian is defined by LAD of *Inoceramus (Cladoceramus) undulatoplicatus*, and the base of the Upper Santonian is defined by the FAD of the crinoid *Uintacrinus socialis*.

It is noteworthy that the generic concept of inoceramid bivalves is used inconsistently. Some authors only recognize one broad genus, *Inoceramus*, which includes several subgenera, whereas others recognize several genera. In accordance with Tröger (1989), we assign all species to *Inoceramus* and also include subgeneric names.

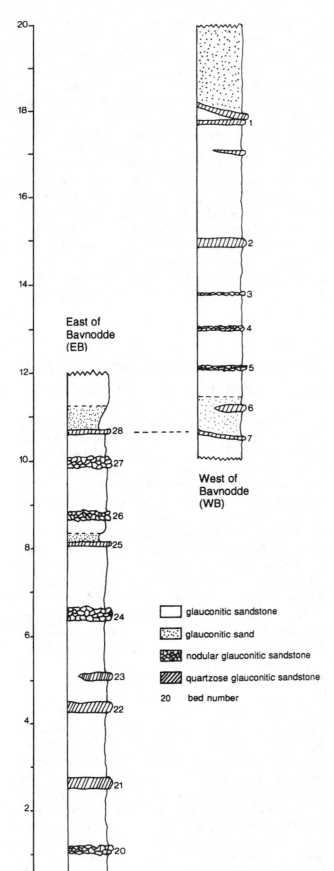

East of
Bavnodde
(EB)

West of
Bavnodde
(WB)

☐ glauconitic sandstone

▨ glauconitic sand

▨ nodular glauconitic sandstone

▨ quartzose glauconitic sandstone

20   bed number

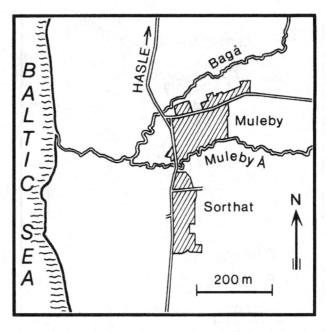

Fig. 7. Map showing the brooks Bagå and Muleby Å close to the west coast of Bornholm. Exposures showing the Arnager Limestone Formation were recorded from stream cuttings of the Muleby Å and the abandoned marl pit immediately west of Sorthat–Muleby (for details, see Fig. 8). After Tröger & Christensen (1991).

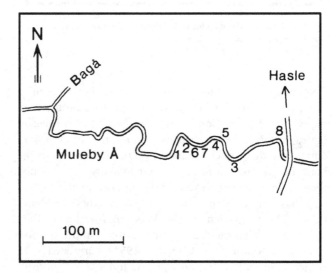

Fig. 8. Map showing outcrops along the brook Muleby Å (after Ravn 1946). *I. (Cremnoceramus) schloenbachi* is recorded from loc. 1, and *Goniocamax lundgreni* occurs at loc. 1, 3, 5, and 8. Loc. 8 is the abandoned marl pit yielding the so-called 'Glass-marl' of Ravn (1946) and Birkelund (1957).

Fig. 6. Sections of the Bavnodde Greensand exposed west and east of Bavnodde, showing approximately the upper 20 m of the greensand. Bed 7 of the section west of Bavnodde probably equates with bed 28 of the section east of Bavnodde.

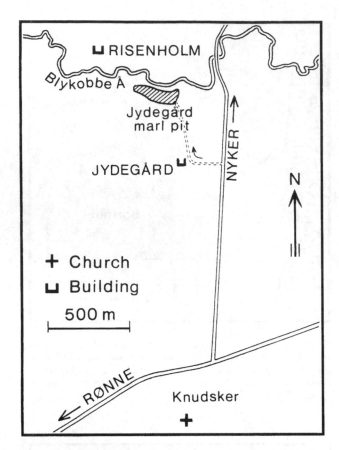

Fig. 9. Map showing the abandoned Jydegård marl pit and Blykobbe Å at Risenholm. After Kennedy & Christensen (1991).

Kennedy (1984) subdivided the Coniacian of Aquitaine, the type region of the stage, into four ammonite zones: the Lower Coniacian *Forresteria petrocoriensis* Zone, the Middle Coniacian *Peroniceras tridorsatum* Zone, and the Upper Coniacian *Gauthiericeras margae* Zone below and *Paratexanites serratomarginatus* Zone above (Fig. 10). The inoceramid sequence of the Aquitaine has not yet been worked out (Kennedy & Cobban 1991). The Coniacian of the Western Interior of the United States has yielded both ammonites and inoceramids, and Kennedy & Cobban (1991) correlated the French ammonite zonation with the inoceramid succession in the Western Interior (Fig. 10).

Tröger (1989) reviewed the inoceramid zonation of the Upper Cretaceous of the North European Province *sensu* Kauffman (1973), which extends from Ireland in the west to the Ural Mountains in the east, and recognized 10 inoceramid assemblage zones in the Coniacian and Santonian (Fig. 10). Walaszczyk (1992) proposed an inoceramid zonation of the Turonian–Santonian of Poland, including nine zones in the Coniacian and Santonian (Fig. 10).

The ammonite and inoceramid zonations are tentatively correlated in Fig. 10. There is no general agreement on the subdivison of the Coniacian and Santonian. Moreover, the taxonomy of many inoceramids is in a state of flux, and the species concept differs from one author to another. For example, *I. frechi sensu* Kennedy & Cobban (1991), which bridges the Upper Turonian and Lower Coniacian in the Western Interior, is not conspecific with *I. frechi sensu* Tröger (1989) and Walaszczyk (1992). *I. frechi sensu* Tröger occurs in his inoceramid assemblage zone 22, while Walaszczyk (1992) records this taxon from his *Cremnoceramus deformis* Zone. According to Walaszczyk (1992), *I. (Cremnoceramus) crassus* is conspecific with *I. (C.) schloenbachi*, and *I. (C.) brongniarti* is conspecific with *I. (C.) erectus* and *I. (C.) rotundatus*. Ravn (1921, Pl. 2:2) described *Inoceramus cordiformis* from the Bavnodde Greensand at Horsemyre Odde. Seitz (1961, p. 114) placed this specimen in *I. (Cordiceramus) cordiformis cordiformis*, whereas Tröger *in* Tröger & Christensen (1991, p. 33) assigned it to *I. (Sphenoceramus) cardissoides* subsp. indet.

Coniacian and Santonian belemnite and inoceramid stratigraphy of Lägerdorf

Ernst & Schulz (1974) (see also Schulz *et al.* 1984; Schulz, 1996) established a local zonation of the upper Coniacian and Santonian of Lägerdorf, NW Germany, on the basis of species of the genus *Gonioteuthis*, inoceramid bivalves, echinoids, and planctic crinoids (Fig. 11). This zonation is discussed below, because it is used as a stratigraphical framework in the present paper, and the zonation is tentatively correlated with the inoceramid assemblage zones of the North Temperate Province of Tröger (1989) (Fig. 11). Moreover, the evolution of the *Gonioteuthis* lineage in the upper Coniacian and Santonian has been studied in detail only at Lägerdorf (Ernst & Schulz 1974).

The *G. praewestfalica* Zone equates with the the upper part of the Middle Coniacian *involutus/bucailli* Zone and the Upper Coniacian *bucailli/praewestfalica* Zone. Bailey *et al.* (1984) suggested that the *bucailli/praewestfalica* Zone should be attributed to a low level in the Santonian, because they recorded the foraminifer *Stensioeina granulata polonica* from this zone at Lägerdorf. *Stensioeina praeexsculpta* 'Form f' (= *S. granulata polonica*) was earlier thought to enter at the base of the *pachti/undulatoplicatus* Zone and thus to mark the Coniacian–Santonian boundary (Koch *in* Ernst & Schulz 1974). Schönfeld (1990, p. 137) showed, however, that *S. granulata polonica* enters in the lower Middle Coniacian *koeneni* Zone of Lägerdorf. The entry point of *S. granulata polonica*, therefore, cannot be used as a marker for the base of the Santonian. *I. (Volviceramus) involutus* is rather common in the lower half of the *bucailli/praewestfalica* Zone, implying that this zone is Coniacian.

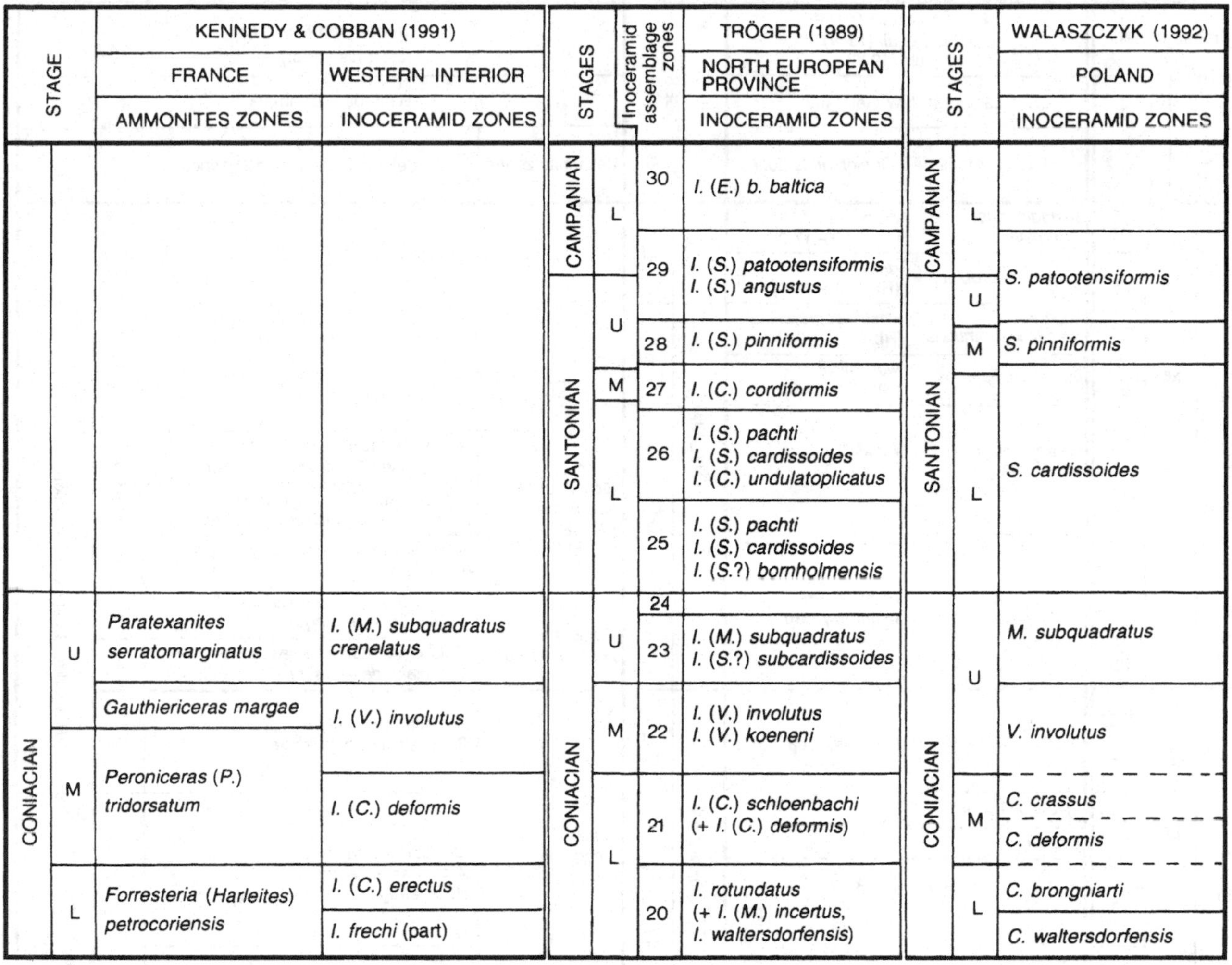

| STAGE | KENNEDY & COBBAN (1991) | | STAGES | Inoceramid assemblage zones | TRÖGER (1989) NORTH EUROPEAN PROVINCE | STAGES | WALASZCZYK (1992) POLAND |
	FRANCE AMMONITES ZONES	WESTERN INTERIOR INOCERAMID ZONES			INOCERAMID ZONES		INOCERAMID ZONES
			CAMPANIAN L	30	I. (E.) b. baltica	CAMPANIAN L	
			SANTONIAN U	29	I. (S.) patootensiformis / I. (S.) angustus	SANTONIAN U	S. patootensiformis
				28	I. (S.) pinniformis	(M)	S. pinniformis
			M	27	I. (C.) cordiformis		
			L	26	I. (S.) pachti / I. (S.) cardissoides / I. (C.) undulatoplicatus	L	S. cardissoides
				25	I. (S.) pachti / I. (S.) cardissoides / I. (S.?) bornholmensis		
			CONIACIAN U	24		CONIACIAN U	
CONIACIAN U	Paratexanites serratomarginatus	I. (M.) subquadratus crenelatus		23	I. (M.) subquadratus / I. (S.?) subcardissoides		M. subquadratus
M	Gauthiericeras margae	I. (V.) involutus	M	22	I. (V.) involutus / I. (V.) koeneni		V. involutus
	Peroniceras (P.) tridorsatum	I. (C.) deformis	L	21	I. (C.) schloenbachi (+ I. (C.) deformis)	M	C. crassus
							C. deformis
L	Forresteria (Harleites) petrocoriensis	I. (C.) erectus		20	I. rotundatus (+ I. (M.) incertus, I. waltersdorfensis)	L	C. brongniarti
		I. frechi (part)					C. waltersdorfensis

Fig. 10. Stratigraphical correlation diagram of the Coniacian – Lower Campanian. *Cremnoceramus crassus* of Walaszczyk (1992)=*I. (C.) schloenbachi* of authors; *C. brongniarti* of Walaszczyk=*I. (C.) erectus* and *I. (C.) rotundatus* of authors. Inoceramid assemblage zone 24 of Tröger (1989) is characterized by *I. (Magadiceramus) subquadratus* and the absence of involute inoceramids.

Gonioteuthis is virtually absent in the Lower Santonian *pachti/undulatoplicatus* Zone of Lägerdorf. Ernst & Schulz (1974, Fig. 15) recorded only one specimen from the top of the zone and a fragment from the middle part of the zone. The transition from *G. praewestfalica* to *G. westfalica* is thus not documented, and the boundary between the *praewestfalica* and *westfalica* Zones has not been established at Lägerdorf.

The lower *westfalica* Zone roughly equates with the *coranguinum/westfalica* Zone and is characterized by samples of *G. westfalica* with a mean Riedel-Index<8.5. *I. (Sphenoceramus)* ex gr. *pachti/cardissoides* also occurs in this zone, which was placed in the upper Lower Santonian by Ernst & Schulz (1974).

The upper *westfalica* Zone equates with the *rogalae/westfalica* Zone and is defined by samples of *G. westfalica* with mean Riedel-Indices of 8.5–11.0. Ernst &

Schulz (1974) placed this zone in the lower Middle Santonian.

I. (Cordiceramus) cordiformis is rather common at the base of the *rogalae/westfalica* Zone, and this flood occurrence probably corresponds to the upper part of the *cordiformis* beds in other sections (Ernst 1966; Ernst & Schulz 1974), i.e. the middle part of zone 27 of Tröger (1989).

I. (Sphenoceramus) pinniformis occurs very rarely in the upper part of the *rogalae/westfalica* Zone of Lägerdorf (Ernst 1966; Ernst & Schulz 1974). Tröger (1989, Fig. 4) correlated inoceramid assemblage zone 28, which is defined by the total range of *I. (S.) pinniformis*, with the upper part of the *westfalicagranulata* Zone and the superjacent lower *granulata* Zone. This correlation, which we have adopted in Fig. 11, is most likely valid in Westfalia and Lower Saxony (Seitz 1965), in addition to the Subher-

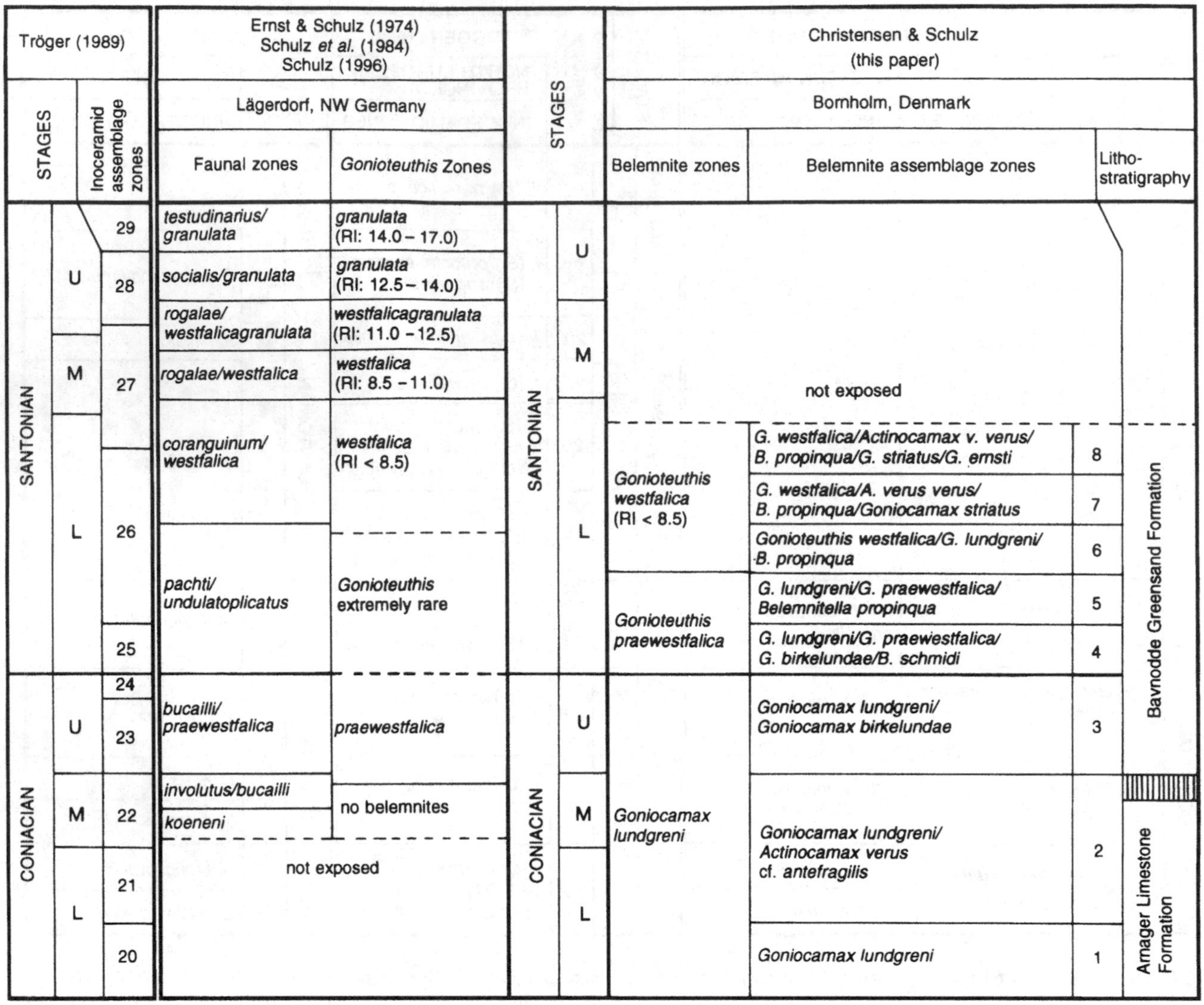

Tröger (1989)			Ernst & Schulz (1974) Schulz et al. (1984) Schulz (1996)		STAGES	Christensen & Schulz (this paper)			
STAGES	Inoceramid assemblage zones		Lägerdorf, NW Germany			Bornholm, Denmark			
			Faunal zones	Gonioteuthis Zones		Belemnite zones	Belemnite assemblage zones		Litho-stratigraphy
SANTONIAN	U	29	testudinarius/ granulata	granulata (RI: 14.0 – 17.0)	SANTONIAN — U	not exposed			Bavnodde Greensand Formation
	U	28	socialis/granulata	granulata (RI: 12.5 – 14.0)					
	M	27	rogalae/ westfalicagranulata	westfalicagranulata (RI: 11.0 – 12.5)	M				
	M	27	rogalae/westfalica	westfalica (RI: 8.5 – 11.0)					
	L	26	coranguinum/ westfalica	westfalica (RI < 8.5)	L	Gonioteuthis westfalica (RI < 8.5)	G. westfalica/Actinocamax v. verus/ B. propinqua/G. striatus/G. ernsti	8	
	L	26					G. westfalica/A. verus verus/ B. propinqua/Goniocamax striatus	7	
	L	26					Gonioteuthis westfalica/G. lundgreni/ B. propinqua	6	
	L	26	pachti/ undulatoplicatus	Gonioteuthis extremely rare		Gonioteuthis praewestfalica	G. lundgreni/G. praewestfalica/ Belemnitella propinqua	5	
	L	25					G. lundgreni/G. praewestfalica/ G. birkelundae/B. schmidi	4	
CONIACIAN	U	24			CONIACIAN — U	Goniocamax lundgreni	Goniocamax lundgreni/ Goniocamax birkelundae	3	
	U	23	bucailli/ praewestfalica	praewestfalica					
	M	22	involutus/bucailli koeneni	no belemnites	M	Goniocamax lundgreni	Goniocamax lundgreni/ Actinocamax verus cf. antefragilis	2	Amager Limestone Formation
	L	21	not exposed		L				
	L	20					Goniocamax lundgreni	1	

Fig. 11. Stratigraphical diagram of the Coniacian and Santonian showing inoceramid assemblage zones of the North European Province (cf. Fig. 10), faunal zones and *Gonioteuthis* Zones of Lägerdorf, and belemnite zones and assemblage zones of Bornholm.

cynian Cretaceous Basin (Ulbrich 1971). However, in other areas, including Lägerdorf, southern England, and Poland, this correlation is very questionable, because *I. (S.) pinniformis* enters earlier. *I. (S.)* ex gr. *pinniformis* appears immediately below Whitaker's 3-inch Flint Band in Kent (Bailey *et al.* 1983), which probably equates with flint F260 of Lägerdorf (Schulz 1996). Flint F260 lies slightly above the base of the lower *westfalica* Zone. *I. (S.) pinniformis* enters approximately one zone later in Poland, i.e. some way above the base of the upper *westfalica* Zone (Walaszczyk 1992). The base of the *I. (S.) pinniformis* Zone is thus diachronous across Europe. The reason for that may be that the index species is facies-controlled or the species concept differs.

The *westfalicagranulata* Zone equates with the *rogalae/ westfalicagranulata* Zone and is characterized by samples of *G. westfalicagranulata* with mean Riedel-Indices of 11.0–12.5. Ernst & Schulz (1974) placed this zone in the upper Middle Santonian. Inoceramids recorded from this zone are *I. (Sphenoceramus) pachti* and *I. (S.)* ex gr. *pachti/ cardissoides. Uintacrinus socialis* enters in the upper part of the zone.

The Upper Santonian is subdivided into two *Gonioteuthis granulata* zones as well as two zones based on the

Fig. 12. Diagram showing the stratigraphical distribution of belemnites and the age of the localities according to the present study.

Range chart showing Lithostratigraphy and bed numbers, Localities (Arnager–Sose block: Arnager – E of Horsemyre Odde, Stampe Å, E of Horsemyre Odde, E of Forchhammers Odde, Forchhammers Odde – Skidteper, W of Skidteper, W of Horsemyre Odde, Horsemyre Odde, W of Forchhammers Odde, Forchhammers Odde, E of Bavnodde, W of Bavnodde; Nyker bl.: Muleby Å, Risenholm, Jydegård), Belemnites (Actinocamax verus cf. antefragilis, Actinocamax v. verus, Gonioteuthis praewestfalica, Gonioteuthis westfalica, Gonioteuthis ernsti sp. nov., Goniocamax (?) sp. nov.(?), Goniocamax lundgreni, Goniocamax sp. nov.(?) aff. lundgreni, Goniocamax birkelundae sp. nov., Goniocamax striatus sp. nov., Belemnitella schmidi sp. nov., Belemnitella propinqua), and Stages (LOWER SANTONIAN: upper, middle, lower; UPPER CONIACIAN; MIDDLE CONIAC.; LOWER CONIACIAN: upper, lower).

planctic crinoids *Uintacrinus socialis* and *Marsupites testudinarius*. The *socialis/granulata* Zone below is characterized by samples of *Gonioteuthis* with mean Riedel-Indices of 12.5–14.0, whereas samples from the *testudinarius/granulata* Zone above have mean Riedel-Indices of 14.0–17.0. *I.* (*Sphenoceramus*) ex gr. *lingua/patootensis* is common in the upper part of the *testudinarius/granulata* Zone (Schulz *et al.* 1984), indicating inoceramid assemblage zone 29, the base of which was correlated with the base of the *M. testudinarius* Zone by Tröger (1989, Fig. 4).

Age of the Arnager Limestone Formation

West of Arnager. – The Arnager Limestone at its type locality west of Arnager has yielded a diverse fauna of inoceramid bivalves, which includes ten species (Tröger & Christensen 1991). Four species bridge the Turonian–Coniacian boundary: *I.* (*Mytiloides*) *incertus, I.* (*Striatoceramus*) *striatoconcentricus, I.* (*Heroceramus*) *hercules,* and *I. lusatiae*. Three species occur in the lower Lower Coniacian inoceramid assemblage zone 20 and the lower half of zone 21 of Tröger (1989): *I.* (*Cremnoceramus*) *waltersdorfensis hannovrensis, I.* cf. *rotundatus,* which is the index species of zone 20, and *I.* cf. *wandereri*. The latter is an east-European species that is very rare in western Europe.

The sum of evidence suggests that the limestone exposed immediately west of Arnager can be referred to the lower Lower Coniacian inoceramid assemblage zone 20 (Figs. 10–12).

The very top of the formation exposed east of Horsemyre Odde yielded *I.* cf. *lusatiae, I.* (*Heroceramus*) cf. *hercules,* and *I.* (*Sphenoceramus*) cf. *guerichi* and thus belongs to the same zone as the fauna from the limestone exposed immediately west of Arnager, i.e. the lower Lower Coniacian inoceramid assemblage zone 20 of Tröger (1989).

Ammonites are rare in the Arnager Limestone, and five species were recorded by Kennedy & Christensen (1991), including the Middle Coniacian index ammonite, *Peroniceras tridorsatum,* and *Scaphites kieslingswaldensis kieslingswaldensis*. One specimen of *P. tridorsatum* was described by Kennedy & Christensen (1991), and another and better preserved specimen was collected loose recently. This specimen was kindly determined by Dr. W.J. Kennedy, Oxford, and is shown here in Fig. 13. Precisely located specimens of *S. k. kieslingswaldensis* are from 2 m above the base and the very top of the formation exposed east of Horsemyre Odde. This subspecies appears in Europe in the Lower Coniacian *I. erectus* Zone and continues into the Upper Coniacian (Kaplan *et al.* 1987).

It is enigmatic that the Middle Coniacian index ammonite *P. tridorsatum* occurs together with inoceramids that are typical for inoceramid assemblage zone 20 in the Arnager Limestone. This discrepancy cannot be explained at present. It is not a matter of definition of the Coniacian

Fig. 13. Peroniceras tridorsatum (Schlüter), MGUH 23739, from the Arnager Limestone. It was collected loose west of Arnager. ×1.

substages, because the inoceramids occur earlier than *P. tridorsatum* elsewhere.

Schiøler (1992) listed the very rich dinoflagellate cyst assemblage from the type locality and noted that key taxa of this assemblage indicate an Early to mid-Coniacian age. Bailey & Hart (1979), on foraminiferal evidence, placed the limestone in the lower part of the Coniacian. Packer & Hart (1994) referred the limestone to the mid-Coniacian.

The limestone at the southwest coast has yielded *Goniocamax l. lundgreni,* with precisely located specimens from the bottom conglomerate, 0.5 m above the base, 1.2 m above the base, 1.5–2.5 m above the base, and the very top of the formation, in addition to *Goniocamax*(?) sp.nov.(?) from the bottom conglomerate.

We consider the limestone at its type locality as lower Lower Coniacian on the basis of its diverse inoceramid fauna.

Stampe Å. – The glauconitic, sandy marl of the upper part of the Arnager Limestone Formation at loc. 5 at the brook Stampe Å has yielded the index species of inoceramid assemblage zone 22, *I.* (*Volviceramus*) *koeneni,* in addition to *I.* (*V.*) *alievimussensis* (Figs. 10–11). This locality has also yielded *Goniocamax l. lundgreni* and *Actinocamax verus* cf. *antefragilis*. No ammonites are recorded from this outcrop.

Stolley (1897, Pl. 3:23) recorded *Actinocamax propinquus* mut. (var.) nov. (=*G. lundgreni*; see below) from Stampen. Birkelund (1957) recorded *G. lundgreni lundgreni*? (=*G. lundgreni*; see below) from the calcareous greensand at Stampe Å and *G. lundgreni excavata*? (=*G. lundgreni*; see below) from the greensand at Stampen. She tentatively regarded the calcareous greensand at Stampe Å as a lateral equivalent of the Arnager Limestone Formation, while the greensand at Stampen was placed in the Bavnodde Greensand Formation. Christensen *in* Tröger & Christensen (1991) suggested that the locality names 'Stampen' and 'Stampe Å' were used synonymously by earlier authors and that the belemnites recorded by them most likely came from loc. 5.

As shown below, the sample of *G. lundgreni* from Stampe Å is more advanced than the sample of *G. lundgreni* from the Arnager Limestone west of Arnager with respect to the depth of the pseudoalveolus and slenderness. On inoceramid and belemnite evidence, therefore, the sandy, glauconitic marl at Stampe Å (loc. 5) is younger than the Arnager Limestone at the type locality, i.e. lower Middle Coniacian (Figs. 10–12).

Muleby Å. – The calcareous sands and marls of the Arnager Limestone Formation, exposed at stream cuttings along the brook Muleby Å and the abandoned marl pit at Muleby (Figs. 7–8), have yielded *Goniocamax lundgreni* (loc. 1, 3, 5, and 8), *Goniocamax* sp.nov.(?) aff. *lundgreni* (loc. 5), and *Actinocamax verus* cf. *antefragilis*. In addition, *I. (Cremnoceramus) schloenbachi* was recorded from loc. 1 by Tröger & Christensen (1991). This index species characterizes inoceramid assemblage zone 21 of Tröger (1989) (Figs. 10–11). The other outcrops have yielded neither inoceramids nor ammonites. Ravn (1946) and Birkelund (1957) regarded the marl at Muleby to be a contemporaneous latereal equivalent of the Arnager Limestone on belemnite evidence.

As shown below the sample of *G. lundgreni* from Muleby Å is slightly more advanced than the sample of *G. lundgreni* from the at Arnager Limestone west of Arnager with respect to the depth of the pseudoalveolus and slenderness. On the basis of inoceramids and belemnites the deposits at Muleby Å, therefore, are younger than the limestone at Arnager and older than the glauconitic, sandy marl at Stampe Å (Fig. 12).

Age of the Bavnodde Greensand Formation

Ammonites and inoceramid bivalves are rare in the greensand and have not been recorded from all outcrops. Belemnites are more common, however, and have been collected from all sites.

Kennedy & Christensen (1991) recorded seven ammonites from the greensand, including the age diagnostic *Scaphites kieslingswaldensis fischeri*, *Texanites pseudotexanum*, and *Baculites* cf. *fuchsi*. *S. kieslingswaldensis fischeri* is the lineal descendant of *S. kieslingswaldensis kieslingswaldensis* and occurs from the Lower Santonian to the lowest Campanian. *B. fuchsi* and *T. pseudotexanum* are Santonian.

Tröger & Christensen (1991) recorded four inoceramid bivalve species from the greensand: *I. (Sphenoceramus) subcardissioides*, *I. (S.?) bornholmensis*, *I. (S.) pachti*, and *I. (S.) cardissoides*, in addition to *I. (Magadiceramus)* ex gr. *subquadratus*.

Birkelund (1957) described six belemnite taxa from the greensand: *Gonioteuthis westfalica*, *G.* aff. *westfalica*, *G.* sp. transitional between *G. westfalica* and *G. granulata*, *Goniocamax lundgreni excavata*, *Actinocamax verus*, and *Belemnitella propinqua ravni*. The specimens of *G. westfalica*, *G.* aff. *westfalica*, and *G. lundgreni excavata* from Jydegård, Risenholm, Horsemyre Odde, and the district between east of Forchhammers Odde and Horsemyre Odde are here assigned to *Goniocamax l. lundgreni*. The specimens of *G.* sp. transitional between *G. westfalica* and *G. granulata* from Risenholm are assigned to *Goniocamax birkelundae* sp.nov. Christensen (1971) placed *Belemnitella propinqua ravni* in synonymy with *B. p. propinqua*.

Birkelund (1957, Table 1) placed the Bavnodde Greensand on the southwest coast in the *I. cordiformis* Zone, which she considered to be upper Lower Santonian. According to Tröger (1989) this zone spans the uppermost Lower, Middle, and basal Upper Santonian (Fig. 10). Birkelund considered the greensand at Jydegård to be slightly older than that at Risenholm and on the southwest coast.

Packer & Hart (1994) placed the Bavnodde Greensand in the Early(?)–Middle Santonian and suggested the presence of a hiatus, including the Upper Coniacian – Lower Santonian, between the Arnager Limestone and Bavnodde Greensand. This is at variance with the results presented here (see below).

The age of the outcrops based on the belemnite assemblages as well as other evidence is discussed below, in ascending stratigraphical order (Fig. 12).

East of Forchhammers Odde, between Forchhammers Odde and Skidteper, west of Skidteper, east of Horsemyre Odde, and west of Horsemyre Odde. – Because of the presumed strike and dip of the greensand, the two faults (see above), and the course of the coastline, the exposures east of Horsemyre Odde, east of Forchhammers Odde, between Forchhammers Odde and Skidteper, and west of Skidteper exhibit the oldest greensand, whereas the greensand exposed west of Horsemyre Odde is slightly younger (Figs. 5, 12). The belemnite assemblages from these outcrops consist virtually of *Goniocamax lundgreni*. In addition to this species, we have seen only six specimens of *Goniocamax birkelundae* sp.nov., which came from the basal part of the section exposed east of Forchhammers Odde (Table 1).

We have at hand 17 belemnites from the outcrop between Forchhammers Odde and Skidteper, 21 specimens from west of Skidteper, seven specimens from east of Horsemyre Odde, and three specimens from west of Horsemyre Odde. The reason that *G. birkelundae* sp.nov. is recorded only from east of Forchhammers Odde is most likely the small number of specimens from the other outcrops. Neither ammonites nor inoceramid bivalves are recorded from the outcrops. The greensand at these exposures is considered to be Upper Coniacian because it is younger than the Arnager Limestone Formation (Lower to Middle Coniacian) and older than the greensand at Jydegård and Horsemyre Odde, where *Gonioteuthis praewestfalica* occurs together with Lower Santonian inoceramid bivalves (Fig. 12).

Blykobbe Å at Risenholm. – We have seen only nine belemnites from this locality, and they are assigned to *Goniocamax l. lundgreni* (six specimens) and *G. birkelundae* sp.nov. (three specimens).

Tröger & Christensen (1991) placed this locality in the lower part of inoceramid assemblage zone 25, i.e. basal Lower Santonian, on inoceramid bivalve evidence (Fig. 10). However, *I. (Magadiceramus)* ex gr. *subquadratus* and *I. (Sphenoceramus) subcardissoides*, which are Upper Coniacian, are also recorded from this outcrop (Tröger & Christensen 1991, p. 35). Kennedy & Christensen (1991) reported *Scaphites kieslingswaldensis fischeri* and *Baculites* cf. *fuchsi*, which are Santonian, from Risenholm. We consider the Bavnodde Greensand Formation exposed here to be uppermost Coniacian and basal Santonian (Fig. 12).

Jydegård. – The belemnite assemblage from this outcrop consists of *Goniocamax l. lundgreni* with subordinate *Goniocamax birkelundae* sp.nov., *Gonioteuthis praewestfalica*, and *Belemnitella schmidi* sp.nov. (Table 2).

Tröger & Christensen (1991) placed the greensand at Jydegård in the uppermost part of inoceramid assemblage zone 24 and the lowest part of inoceramid assemblage zone 25, i.e. uppermost Coniacian and lowest Santonian, on inoceramid bivalve evidence (Fig. 10). No age diagnostic ammonites are recorded from Jydegård.

On the basis of the belemnite assemblage the greensand at Jydegård is considered to be slightly younger than that of Risenholm, since *G. praewestfalica* and *B. schmidi* sp.nov. occur at Jydegård, and *G. birkelundae* sp.nov. is less common at Jydegård than at Risenholm. The greensand at Jydegård is basal Santonian (Fig. 12) but may bridge the Coniacian–Santonian boundary, because *G. lundgreni* predominates, *G. praewestfalica* is upper Middle to Upper Coniacian elsewhere, and *B. schmidi* is most likely the ancestor of *B. p. propinqua*.

Table 1. Estimate of the relative abundance of belemnites from the Bavnodde Greensand, east of Forchhammers Odde.

Goniocamax lundgreni	76	(93%)
Goniocamax birkelundae sp.nov.	6	(7%)
Total	82	

Table 2. Estimate of the relative abundance of belemnites from the Bavnodde Greensand, Jydegård.

Goniocamax lundgreni	74	(85%)
Goniocamax birkelundae sp.nov.	3	(4%)
Gonioteuthis praewestfalica	8	(9%)
Belemnitella schmidi sp.nov.	2	(2%)
Total	87	

Table 3. Estimate of the relative abundance of belemnites from the Bavnodde Greensand, Horsemyre Odde.

Goniocamax lundgreni	45	(83%)
Gonioteuthis praewestfalica	7	(13%)
Belemnitella propinqua	2	(4%)
Total	54	

Table 4. Estimate of the relative abundance of belemnites from the Bavnodde Greensand, west of Forchhammers Odde.

Gonioteuthis westfalica	35	(95%)
Goniocamax lundgreni	2	(5%)
Total	37	

Horsemyre Odde. – The belemnite assemblage from this outcrop consists of *Goniocamax l. lundgreni* with subordinate *Gonioteuthis praewestfalica*, and *Belemmnitella p. propinqua* (Table 3). Kennedy & Christensen (1991) recorded *Scaphites kieslingswaldensis fischeri*, and Tröger & Christensen (1991) reported *I. (Sphenoceramus) cardissoides* subsp. indet. from this outcrop.

The greensand at Horsemyre Odde is younger than that exposed to the east and west owing to the strike and dip of the sediments, the course of the coast line, and the fault immediately east of Horsemyre Odde. According to observations made by the authors in 1993, there is a gap of 6–8 m of unexposed sediments between the outcrop west of Horsemyre Odde and Horsemyre Odde (Fig. 5).

The belemnite assemblage from Horsemyre Odde is younger than that from Jydegård, because it includes *B. p. propinqua*, which is considered the lineal descendant of *B. schmidi* sp.nov. We therefore place the greensand at this site in the middle part of the lower Lower Santonian on belemnite, ammonite, and inoceramid bivalve evidence (Fig. 12). This is at variance with the result of Packer *et al.*

Table 5. Estimate of the relative abundance of belemnites from the Bavnodde Greensand, Forchhammers Odde.

Gonioteuthis westfalica	31	(91%)
Goniocamax lundgreni	3	(9%)
Total	34	

Table 6. Estimate of the relative abundance of belemnites from the Bavnodde Greensand, east of Bavnodde.

Gonioteuthis westfalica	97	(90%)
Actinocamax verus	6	(6%)
Goniocamax striatus sp.nov.	1	(1%)
Belemnitella propinqua	3	(3%)
Total	107	

Table 7. Estimate of the relative abundance of belemnites from the Bavnodde Greensand, west of Bavnodde.

Gonioteuthis westfalica	418	(90%)
Gonioteuthis ernsti sp.nov.	6	(1%)
Actinocamax verus	22	(5%)
Goniocamax striatus sp.nov.	6	(1%)
Belemnitella propinqua	12	(3%)
Total	464	

(1989, Fig. 6), who placed the greensand at Horsemyre Odde in the Upper Coniacian on foraminiferal evidence.

West of Forchhammers Odde and Forchhammers Odde. – We have collected *Gonioteuthis w. westfalica* (common) and *Goniocamax l. lundgreni* (very rare) from these outcrops (Tables 4–5). The age of these localities is discussed below.

According to earlier records, *Belemnitella propinqua* occurs at Forchhammers Odde and Forchhammers Klint. The latter locality includes east of Bavnodde, west of Forchhammers Odde, and Forchhammers Odde as used herein (see above).

East of Bavnodde. – The belemnite assemblage includes *Gonioteuthis w. westfalica* with subordinate *Actinocamax v. verus*, *Goniocamax striatus* sp.nov., and *Belemnitella p. propinqua* (Table 6). The age of this outcrop is discussed below.

West of Bavnodde. – The belemnite assemblage from this locality is almost identical to that of east of Bavnodde (Table 7), but in addition *Gonioteuthis ernsti* sp.nov. occurs in the lower third of the section (Fig. 12). The age of this outcrop is discussed below.

Discussion of the age of the outcrops between west of Bavnodde and Forchhammers Odde. – Tröger & Christensen (1991) recorded *I. (Sphenoceramus) pachti* and *I. (S.) cardissoides* from west of Bavnodde, including precisely

located specimens from between beds 5 and 6. The greensand at Bavnodde was placed in the middle and upper part of inoceramid assemblage zone 25 and the lowest part of zone 26, i.e. lower Lower Santonian, on inoceramid bivalve evidence (Fig. 10).

Most of the ammonites recorded by Kennnedy & Christensen (1991) are from old collections, and they are therefore labelled only with locality name. *Texanites pseudotexanum* and *Scaphites kieslingswaldensis fischeri* were reported from Bavnodde and Forchhammers Odde, and precisely located specimens of *S. kieslingswaldensis fischeri* are from west of Bavnodde, bed 4 and between beds 3 and 4, and east of Bavnodde, bed 20, between beds 20 and 22, and between 27 and 28.

The belemnite assemblages from west and east of Bavnodde, west of Forchhammers Odde, and Forchhammers Odde are younger than the belemnite assemblage from Horsemyre Odde, because *G. w. westfalica* predominates and constitutes ca. 90–95% of the assemblage. On the basis of the mean Riedel-Index of the samples from east of west of Bavnodde, which is less than 8.5, these localities are placed in the lower *westfalica* Zone *sensu* Ernst & Schulz (1974). Moreover, *G. ernsti* sp.nov., which occurs at west of Bavnodde, between beds 3 and 6, is also known from the Münster Basin, lower *westfalica* beds (=*undulatoplicatus/cordiformis* boundary beds (Ernst 1964), i.e. upper Lower Santonian.

On the basis of the strike of the greensand exposed east and west of Bavnodde, we suggest that: (1) bed 28 of the outcrop east of Bavnodde may equate with bed 7 of the exposure west of Bavnodde (Fig. 6), (2) the exposure west of Forchhammers Odde is slightly older than the exposure at Forchhammers Odde, and (3) the outcrops at Forchhammers Odde are slightly older than the exposure east of Bavnodde. This is consistent with the ranges of the six belemnite species occurring at these localities (Fig. 12).

The specimens of *G. westfalica* from west of Forchhammers Odde and Forchhammers Odde are somewhat flattened ventrally (mean Flattening-Quotient is 1.01) in contrast to those from east and west of Bavnodde and those from the lower *westfalica* Zone of Lägerdorf (Ernst & Schulz 1974, Fig. 15). Therefore, we consider the specimens from west of Forchhammers Odde and Forchhammers Odde as early forms of *G. westfalica*, which are transitional to *G. praewestfalica*.

In conclusion, we thus place the exposures west of Forchhammers Odde and Forchhammers Odde in the middle Lower Santonian, and the exposures east and west of Bavnodde in the upper Lower Santonian (Fig. 12).

Packer *et al.* (1989, Fig. 6) placed the basal 2 m of the exposure east of Bavnodde in the Upper Coniacian on foraminiferal evidence. This is at variance with the dating based on macrofossils. They placed the remaining part of the greensand exposed east of Bavnodde and the greensand exposed west of Bavnodde in the Santonian.

Belemnite stratigraphy

The Coniacian and Lower Santonian belemnite zonation of Bornholm suggested below is local and based on species from the Central European and Central Russian Subprovinces. Three zones and eight assemblage zones are recognized (Fig. 11). The base of most of these zones cannot be defined precisely, because of minor stratigraphical gaps between the exposures (Figs. 11–12). Moreover, it is very likely that the zonation is valid only for the Baltoscandian area for the following reasons: (1) *Goniocamax l. lundgreni* and *Belemnitella p. propinqua* are extremely rare in NW Europe; (2) *Gonioteuthis praewestfalica* enters earlier in NW Germany and southern England; (3) *G. praewestfalica* and *G. ernsti* sp.nov. are not recorded from the Russian Platform, and *G. w. westfalica* is rare on the Russian Platform and not recorded east of Ukraine; and (4) *Goniocamax birkelundae* sp.nov. and *G. striatus* sp.nov. are not recorded outside Bornholm, whereas *Belemnitella schmidi* sp.nov. is known only from Bornholm and the Russian Platform. The zonation is correlated with the zonations of Lägerdorf and the inoceramid assemblage zones of Tröger (1989) (Fig. 11). The zones are listed in ascending stratigraphical order:

1   The *Goniocamax lundgreni* Zone comprises the Coniacian and includes the Arnager Limestone Formation and the lower third of the Bavnodde Greensand Formation. This zone equates partly with the Russian *G. lundgreni* Zone, which is Middle and Upper Coniacian. The base of the zone is defined by the first occurrence of *G. lundgreni* in the conglomerate at the base of the Arnager Limestone. The first occurrence level of the species on Bornholm probably represents its absolute first occurrence level for the following reasons: (1) It is not known from the Turonian elsewhere, (2) it is more advanced than belemnites occurring in the Upper Turonian, and (3) it first appears on the Russian Platform in the Middle Coniacian.

   Three assemblage zones are recognized. Assemblage zone 1 with *G. lundgreni* is lower Lower Coniacian and includes the Arnager Limestone at the type locality; assemblage zone 2 with *G. l. lundgreni* and *Actinocamax verus* cf. *antefragilis* is upper Lower and Middle Coniacian and comprises the Arnager Limestone Formation exposed at Muleby Å and Stampe Å; and assemblage zone 3 with *G. l. lundgreni* and *G. birkelundae* sp.nov. is Upper Coniacian and includes the lower third of the Bavnodde Greensand.

2   The *Gonioteuthis praewestfalica* Zone is lower Lower Santonian and includes the middle part of the Bavnodde Greensand. This zone roughly equates with the lower two-thirds of the German *pachti/undulatoplicatus* Zone. Two assemblages zones are recognized: assemblage zone 4 with *G. l. lundgreni*, *G. praewestfalica*, and *Belemnitella schmidi* sp.nov.; and assemblage zone 5 with *G. l. lundgreni*, *G. praewestfalica*, and *B. p. propinqua*.

3   The *Gonioteuthis westfalica* Zone is middle and upper Lower Santonian and includes the upper part of the Bavnodde Greensand. Three assemblage zones are recognized: assemblage zone 6 with *G. w. westfalica*, *G. l. lundgreni*, and *B. p. propinqua*; assemblage zone 7 with *G. w. westfalica*, *A. v. verus*, *Goniocamax striatus* sp.nov., and *B. p. propinqua*; and assemblage zone 8 with *G. w. westfalica*, *A. v. verus*, *B. p. propinqua*, *G. striatus* sp.nov., and *Gonioteuthis ernsti* sp.nov. The base of assemblage zone 8 is defined by the first occurrence of *G. ernsti* sp.nov. in the greensand exposed west of Bavnodde, between beds 5 and 6 (Fig. 6).

   Since the specimens of *G. w. westfalica* from assemblage zone 6 (west of Forchhammers Odde and Forchhammers Odde) are early forms compared to those from the lower *westfalica* Zone of Lägerdorf (see above), zone 6 is correlated with the upper part of the *pachti/undulatoplicatus* Zone, and zones 7 and 8 with the *coranguinum/westfalica* Zone (Fig. 11).

Palaeobiogeography and evolutionary trends

General

The palaeobiogeography of the Upper Cretaceous belemnites of the Northern Hemisphere was discussed by Christensen (1975a, 1976, 1988, 1990b). The North Temperate Realm (=Boreal Realm of authors) includes the North American and North European Provinces and is characterized by the family Belemnitellidae Pavlow. Belemnitellids occur rarely also in the northern part of the Tethyan Realm, which otherwise in the Cenomanian is characterized by the family Belemnopseidae Naef. The belemnopseids became extinct in the Cenomanian, and afterwards a Tethyan Realm cannot be defined on the basis of belemnites, but it can be recognized on the basis of other fossil groups, such as rudists, ammonites and larger foraminifera.

The North European Province extends from Northern Ireland in the west to the Ural Mountains in the east and includes the Central European and Central Russian Subprovinces (Fig. 14). These subprovinces are well defined from the Middle Coniacian to the Lower Campanian and characterized by independently evolving belemnite lineages: the *Gonioteuthis* stock inhabited the Central European Subprovince, and the *Goniocamax–Belemnitella* stock inhabited the Central Russian Subprovince. *Actinocamax verus* and *Belemnellocamax* ex gr. *grossouvrei*

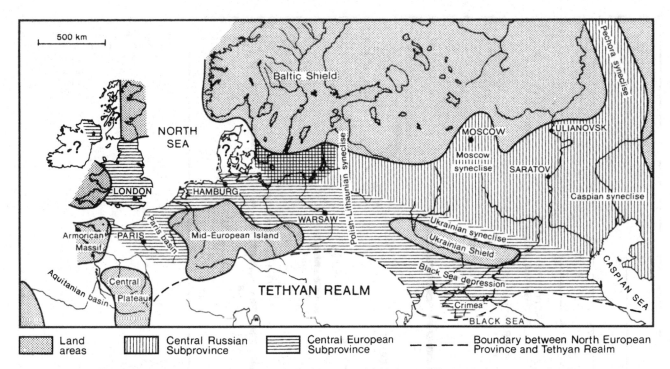

Fig. 14. Distribution of Upper Cretaceous biogeographical units in Europe based on belemnitellids. Upper Cretaceous land and sea areas represent maximum inundation for all stages. The boundaries are not reliable in detail, and the biogeographical units are typically gradational in character. After Christensen (1976).

(Janet) are widely distributed in both subprovinces; *A. verus* occurs commmonly and *B.* ex gr. *grossouvrei* is very rare. In other periods of the Upper Cretaceous the subprovinces are less distinct and may disappear completely.

The Middle Coniacian – Lower Campanian belemnite faunas of Baltoscandia are of great importance, because they include species of the *Gonioteuthis* and *Goniocamax–Belemnitella* stocks and thus provide a basis for correlation between the two subprovinces.

Belemnites are virtually unknown in the upper Turonian and Lower Coniacian of the Central European Subprovince. The small fauna, consisting of about a dozen specimens, was revised by Christensen (1982), who recognized four species, *Actinocamax bohemicus* Stolley, *A. paderbornensis* Schlüter, *A. strehlensis* (Fritsch & Schlönbach), and *Goniocamax esseniensis* (Christensen). Belemnites are more common in the Central Russian Subprovince during this time interval, but they have a very restricted distribution, occurring almost exclusively in the Volga district (Naidin 1981). The following species are recorded: *Actinocamax matesovae* (Naidin), *A. medwedicicus* (Naidin), *A. intermedius* Arkhangelsky, *A. coronatus* (Makhlin), *A. planus* (Makhlin), and various subspecies of *A. verus*.

In addition, belemnites are not recorded or are extremely rare in the Middle Coniacian – basal Santonian of England (Christensen 1982, 1991), Northern Ireland (Reid 1971), and Poland (Cieslinski 1963).

Belemnite faunas of the Coniacian – Lower Santonian of Bornholm

Coniacian. – The Coniacian belemnite fauna consists of *Goniocamax l. lundgreni*, *G. birkelundae* sp.nov., *G.* sp.nov.(?) aff. *lundgreni*, *G.*(?) sp.nov.(?), and *Actinocamax verus* cf. *antefragilis* (Figs. 12, 15). *G. l. lundgreni* predominates, and the other taxa are very or extremely rare. *G. l. lundgreni* is rare in the lower Lower Coniacian but becomes more common in younger beds. We have at hand a little less than two scores of specimens from the lower Lower Coniacian Arnager Limestone, half of which were collected recently. The remaining material was collected during the last 100 years.

The Middle and Upper Coniacian of the Central European Subprovince is characterized by *Gonioteuthis praewestfalica*, while the Central Russian Subprovince is characterized by *G. lundgreni*. The Middle and Upper Coniacian chalk of Lägerdorf has yielded only *G. praewestfalica* (Ernst & Schulz 1974), and only one specimen of *G. praewestfalica* is recorded from the Upper Coniacian chalk of southern England (Bailey *et al.* 1983; Christensen 1991). It is tentatively suggested that the

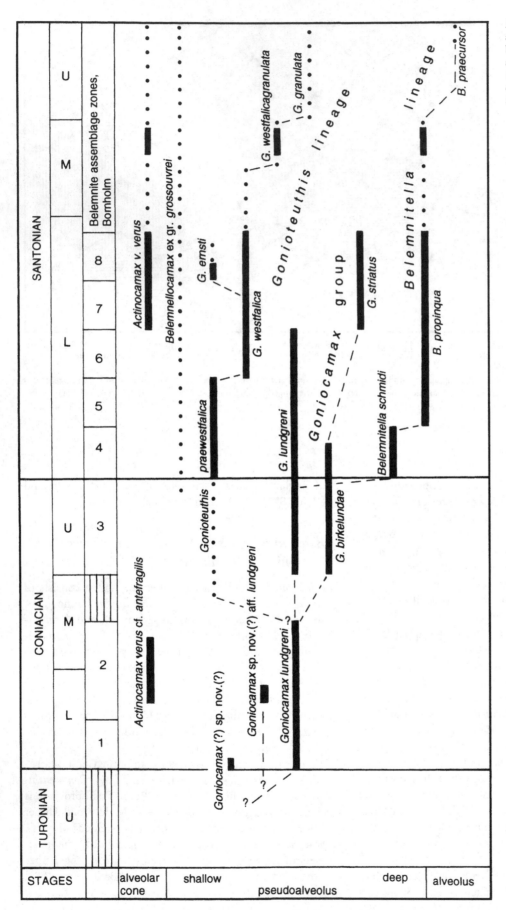

Fig. 15. Phylogenetical relations and stratigraphical ranges of belemnite species in the Coniacian and Santonian of NW Europe (dotted lines), with particular reference to Bornholm and Eriksdal, Scania (thick lines).

Gonioteuthis lineage is derived from *Goniocamax lundgreni* (see below).

The origin of *Goniocamax lundgreni*, which we consider as the earliest member of the genus *Goniocamax*, is unknown. Christensen (1988) suggested that its ancestor should be sought among the Turonian species of *Actinocamax* from the Russian Platform. Naidin (1964b, Fig. 35) derived *G. lundgreni* from *Actinocamax medwedicicus*, but he did not discuss the affinity. Both suggestions are doubtful, because the stratigraphically oldest representatives of the genus *Goniocamax* are known from the lower Lower Coniacian of Bornholm (*G. l. lundgreni*) and the Münster Basin (*G. esseniensis*), whereas *G. l. lundgreni* first appears in the Middle Coniacian in Russia. *G. l. lundgreni* thus appears later on the Russian Platform than on Bornholm, and it is therefore suggested that this taxon migrated to the Russian Platform from Baltoscandia.

Actinocamax verus antefragilis was recorded earlier only from the Lower Turonian of the Russian Platform (Naidin 1964b). The two specimens of *A. verus* cf. *antefragilis* from the Coniacian of Bornholm are much younger. They may be regarded as stray specimens from an unknown area, where the subspecies survived from the Lower Turonian to the Coniacian.

Goniocamax birkelundae sp.nov. is known only from Bornholm, where it appears near the base of the Upper Coniacian. Since it is closely allied to *G. lundgreni*, we suggest that it evolved from *G. lundgreni* elsewhere by allopatric speciation and immigrated later to the Bornholm area.

Santonian. – The Lower Santonian belemnite fauna of Bornholm is extraordinarily diverse and includes nine species: *Actinocamax v. verus*, *Gonioteuthis praewestfalica*, *Gonioteuthis w. westfalica*, *Gonioteuthis ernsti* sp.nov., *Goniocamax l. lundgreni*, *Goniocamax birkelundae* sp.nov., *Goniocamax striatus* sp.nov., *Belemnitella schmidi* sp.nov., and *Belemnitella p. propinqua* (Figs. 12, 15).

The Lower Santonian of the Central European Subprovince is characterized by *Gonioteuthis w. westfalica* and the Central Russian Subprovince by *Belemnitella p. propinqua* and *Goniocamax lundgreni uilicus* (Koltypin) (Naidin & Kopaevich 1977). The Lower Santonian chalk of Lägerdorf has yielded only *G. w. westfalica* (Ernst & Schulz 1974), whereas the Lower Santonian chalk of southern England has yielded a more diverse fauna, including *G. w. westfalica* and *A. v. verus*, with subordinate *G. l. lundgreni*, *Belemnellocamax* ex gr. *grossouvrei*, and *B. p. propinqua* (Christensen 1991).

G. l. lundgreni predominates in the lower Lower Santonian of Bornholm, becomes rare in the middle Lower Santonian, and disappears in the upper Lower Santonian.

G. praewestfalica enters at the base of the Lower Santonian of Bornholm, probably immigrating from the west, and occurs rarely in the lower Lower Santonian. It is fol-

lowed upwards by *G. w. westfalica*, which is most likely the lineal descendant of *G. praewestfalica*. However, the transition is documented neither at Lägerdorf nor on Bornholm. *Gonioteuthis* is extremely rare in the lower Lower Santonian *pachti/undulatoplicatus* Zone of Lägerdorf (Fig. 11), and the boundary beds between the lower and middle Lower Santonian are not exposed on Bornholm (Fig. 12).

G. w. westfalica predominates in the middle and upper Lower Santonian of Bornholm. There is thus a major faunal turnover at the base of the *G. westfalica* Zone of Bornholm, i.e. between belemnite assemblage zones 5 and 6 (Fig. 11). Belemnite assemblages below are dominated by *G. l. lundgreni*, which constitutes about 85–95% of the fauna, and assemblages above are characterized by *G. w. westfalica*, which constitutes about 90–95% of the fauna.

Populations of *Gonioteuthis* from the off-shore white chalk of Lägerdorf only contain adult specimens (Ernst 1964; Ernst & Schulz 1974). In contrast, populations of *G. praewestfalica* and *G. w. westfalica* from the Lower Santonian of Bornholm include all growth stages, indicating that these taxa reproduced in that area.

Gonioteuthis ernsti sp.nov. enters in the uppermost Lower Santonian and occurs rarely on Bornholm and in the Münster Basin. This taxon is closely allied to *G. westfalica* and may have evolved by allopatric speciation from this species (Fig. 15).

Actinocamax v. verus appears near the base of the upper Lower Santonian, i.e. slightly later than *G. w. westfalica*. *Actinocamax* is not recorded from the Upper Coniacian and lower and middle Lower Santonian of Bornholm (Figs. 12, 15). This is probably not due to collection failure, because we have at hand ca. 350 belemnites from beds of this age. The absence of *Actinocamax*, therefore, is most likely real. In the off-shore chalks of NW Germany and southern England, *A. v. verus* is mainly Upper Santonian; it has an acme in the Upper Santonian *Uintacrinus socialis* Zone of Kent (Ernst & Schulz 1974; Bailey *et al.* 1984; Christensen 1991).

Goniocamax striatus sp.nov. enters simultaneously with *A. v. verus* near the base of the upper Lower Santonian (Fig. 12). It is recorded only from Bornholm and may be considered as the lineal descendant of *G. birkelundae* sp.nov. (Fig. 15).

Belemnitella schmidi sp.nov. appears at the base of the Lower Santonian of Bornholm, as does *G. praewestfalica* (Figs. 12, 15). It is also recorded from the Russian Platform. This taxon may have evolved from *G. lundgreni* by allopatric speciation in an unknown area; it later immigrated to Bornholm and the Russian Platform.

Belemnitella p. propinqua appears slightly later than *B. schmidi* sp.nov. (Fig. 12). It is regarded as the lineal descendant of *B. schmidi* sp.nov. (Figs. 12, 15).

In conclusion, Bornholm was part of the Central Russian Subprovince from the Middle Coniacian to the lower

Lower Santonian, because *Goniocamax lundgreni* predominates. During the remaining part of the Lower Santonian, Bornholm was part of the Central European Subprovince, because *Gonioteuthis* prevails with subordinate *Goniocamax* and *Belemnitella*. During the Middle and

Upper Santonian, Baltoscandia was also part of the Central European Subprovince. For instance, *Gonioteuthis westfalicagranulata* predominates with subordinate *Belemnitella propinqua* in the upper Middle Santonian Eriksdal marl in Scania (Christensen 1986) (Fig. 15).

Systematic palaeontology

Morphology of the guard and terminology. – The guard is usually the only part of the skeleton that is preserved in the Upper Cretaceous belemnitellids, and both external and internal characters are used for taxonomical classification (Fig. 16).

The following characters are generally considered to be of taxonomical value in describing Upper Cretaceous belemnitellids: (1) length of guard; (2) shape of guard; (3) structure of anterior end; (4) surface markings; (5) internal characters; and (6) ontogeny. The characters were discussed by Ernst (1964), Christensen (1975a, 1986), and Schulz (1979).

The term conellae is used descriptively for the cone-shaped tubercles that cover the wall of the pseudoalveolus or alveolus of some belemnitellids. Bandel & Spaeth (1988) suggested that the conellae were formed when part of the anterior, aragonitic end of the guard was replaced by calcite during early diagenesis.

Jeletzky (1950, 1961) stressed the taxonomical importance of granulation within the belemnitellids. However, Christensen (1982) did not follow this point of view and showed that too much importance has been attributed to this character. Granulation may be diagnostic at the species level, e.g., *Actinocamax verus* Miller. In the evolutionary lineage of *Gonioteuthis*, granulation is a prominent character, except in its earliest members. *G. praewestfalica* is very rarely granulated (less than 10% of the specimens), whereas around 50% of the specimens of the succeeding species, *G. westfalica*, are granulated (see below). Species of the genus *Belemnellocamax* Naidin, 1964, are usually not granulated, but a few granulated specimens of *B.* ex gr. *grossouvrei* (Janet) and *B. mammillatus* (Nilsson) have been recorded (Naidin 1964b; Christensen 1982). Species of the genus *Goniocamax* Naidin, as used herein, are generally not granulated, but Naidin (1964b, p. 135) recorded a single granulated specimen of *G. lundgreni*. Christensen (1993b) established a new species, *Actinocamax cobbani*, on the basis of ca. 150 specimens and showed that one, possibly two, specimens were granulated. Christensen & Hoch (1983) described a granulated specimen of *A.* cf. *manitobensis* (Whiteaves). This species is otherwise not granulated. It can thus be concluded that granulation may be diagnostic at the species and generic level and that granulation may occur very rarely in genera that usually are not granulated.

Measurements, ratios, and abbreviations. – A list of measured characters and abbreviations is given in Fig. 16. In addition, the following abbreviations are in the descriptions: LDP – lateral diameter at protoconch; MLD – maximum lateral diameter; MDVD – dorsoventral diameter at the same place; and LVF – length of ventral fissure. Lin-

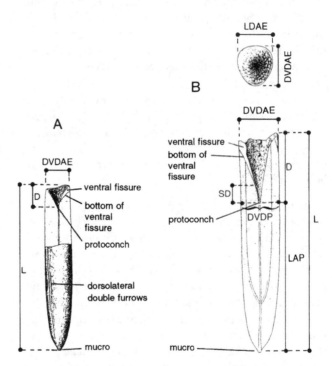

Fig. 16. Morphological elements of the guard in *Goniocamax lundgreni* (A) and *Belemnitella propinqua* (B). L=length of the guard; LAP=length from apex to protoconch; D=depth of pseudoalveolus; SD=Schatzky distance; DVDP=dorsoventral diameter at protoconch; DVDAE= dorsoventral diameter at alveolar end; LDAE=lateral diameter at alveolar end. The alveolar angle (AA) is the angle of the alveolus measured in the median plane; it is measured in the anterior part of the alveolus. The fissure angle (FA) is the angle between the wall of the alveolus and the straight line that connects the intersection points of the bottom of the ventral fissure on the wall of the alveolus and the outer margin of the guard. The Schatzky distance is the distance from the anterior part of the protoconch to the beginning of the bottom of the ventral fissure measured along the longidudinal axis of the guard. The wall of the alveolus of *B. propinqua* is covered by conellae. □A. MGUH 7840 (figured by Birkelund, 1957, Pl. 1:8), Bavnodde Greensand, Jydegård. □B. MGUH 7845 (figured by Birkelund, 1957, Pl. 2:5), Bavnodde Greensand, Forchhammers Klint. Modified from Birkelund (1957). ×0.6.

ear measurements were made with a vernier caliper to an accuracy of 0.1 mm, and angles were measured with a goniometer ocular fitted on a Wild stereomicroscope to an accuracy of 0.5°.

The Birkelund Index (BI) of Christensen (1995) is the ratio of the length from the apex to the protoconch divided by the dorsoventral diameter at the protoconch; the Riedel-Quotient (RQ) of Ernst (1964) is the ratio of the length of the guard divided by the depth of the pseudoalveolus; the 'Schlankheitsquotient' (=Slenderness-Quotient, SQ) of Ernst (1964) is the ratio of the length of the guard divided by the dorsoventral diameter at the alveolar end; the 'Abplattungsquotient' (=Flattening-Quotient, FQ) of Ernst & Schulz (1974) is the ratio of the maximum lateral diameter divided by the dorsoventral diameter at the same place. The term 'Ventralaspekt' of Ernst & Schulz (1974), which is the maximum lateral diameter divided by the lateral diameter at the alveolar end, is not used in the present paper, because this ratio was defined in another way by Schulz (1979).

Ernst & Schulz (1974, p. 45) discussed the mathematical disadvantage of the use of the Riedel-Quotient in species with a very shallow pseudoalveolus. They consequently introduced the Riedel-Index, which is the depth of the pseudoalveolus as a percentage of the length of guard. We follow Ernst & Schulz (1974) and use the Riedel-Index in the present paper. However, we also report the Riedel-Quotient in order to facilitate comparison with samples of belemnites described by earlier authors.

Biometrical methods

Species variation is analyzed using univariate and bivariate statistical methods and is summarized by descriptive statistics, histograms, and scatter diagrams.

Univariate analysis. – The estimates of the following statistics are calculated: arithmetical mean value (X), standard deviation (SD), and coefficient of variation (CV). In addition, the observed range (OR) is reported, and N is the number of specimens. Size-frequency distributions are tested for normality using the powerful non-parametric Kolmogorov–Smirnov test for goodness of fit. This test is discussed by Siegel (1956) and Sokal & Rohlf (1969).

Bivariate analysis. – The regression line is written: $y=a+bx$, and the original measurements are used in the calculations, because of the linear trend on ordinary graph paper and the homoscedastic variance of the regression lines.

In one case the scatter of points is curved, and the growth curve is calculated using the power curve, $y=bx^a$; this curve is also referred to as the equation of simple allometry.

Estimates of the following statistical parameters are calculated: the slope (b), and the standard deviation of the slope (SD_b); the intercept of the y-axis (a), and the standard deviation of the intercept (SD_a); the variance (SD^2_{yx}) and the standard deviation (SD_{yx}) of the regression line; and the correlation coefficient (r). The coefficient of determination, r^2, is the percentage of variation explained by the linear regression. N is the number of specimens. The correlation coefficients are tested for significance by using Table Y of Rohlf & Sokal (1969), and t-tests on the y-intercepts are performed in order to see if the intercepts differ significantly from zero. The latter test has significant biological implications, since a regression line passing through the origin indicates isometric growth, $y=bx$. Regression lines of two samples are compared in the way described by Hald (1957, pp. 571–579).

The relative growth is usually determined adequately by regression lines calculated on the basis of mass-data (Christensen 1975a). It is noteworthy, however, that the growth curves obtained by regression analysis in various species of *Belemnella* differ from the growth curves derived from successive growth stages in specimens ground to or split in the median plane. The growth curves of individual specimens show that the relative growth is strongly allometric, whereas growth curves calculated on the basis of mass-data falsely indicate isometric growth (Christensen 1975a, p. 61; Schulz 1979, p. 39).

Ratios. – Christensen (1973, 1974, 1975a, 1988) and Schulz (1979) discussed the disadvantages of using ratios in palaeontological studies, especially in cases where growth is allometric. In the present paper various ratios are calculated either in order to facilitate comparison with belemnites described by earlier authors, or in those instances where the small number of available specimens prevents bivariate analysis. However, ratios are calculated in those instances only where the relationship of the variates may be regarded as isometric, with a few exceptions.

Order Belemnitida Zittel, 1895
Suborder Belemnopseina Jeletzky, 1965

Family Belemnitellidae Pavlow, 1914

Type genus. – *Belemnitella* d'Orbigny, 1840.

Diagnosis. – See Christensen (1975a).

Distribution. – Belemnitellids are restricted to the Upper Cretaceous and are reported from the Lower Cenomanian to the Upper Maastrichtian. They occur mainly in the North Temperate Realm.

Genus *Actinocamax* Miller, 1823

Type species. – *Actinocamax verus* Miller, 1823, by original designation.

Diagnosis. – Small to large belemnitellids with a conical alveolar fracture of varying length; rarely with a flat anterior end or a very shallow pseudoalveolus; ventral fissure absent; ventral furrow may be present; conical alveolar fracture with concentric growth layers and radiating ribs; guard with dorsolateral longitudinal depressions and double furrows; single lateral furrows sometimes present; vascular imprints usually faint or absent; granulation and striation may be present.

Discussion. – Naidin (1964b) recognized three subgenera of *Actinocamax*: *A. (Actinocamax)*, type species *A. verus* Miller, 1823; *A. (Praeactinocamax)* Naidin, type species *A. plenus* (Blainville, 1825–1827); and *A. (Paractinocamax)* Naidin, type species *A. grossouvrei* Janet, 1891. This classification was discussed by Christensen (1982, 1986, 1991, 1993b, 1994). Christensen (1986, 1991) placed *A. (Paractinocamax)* in synonymy with the genus *Belemnellocamax* Naidin, type species *Belemnites mammillatus* Nilsson, 1826. Naidin (1964b) mainly distinguished *A. (Actinocamax)* and *A. (Praeactinocamax)* by the size of the guard and the length of the cone-shaped alveolar fracture; small species usually with a long cone-shaped alveolar fracture were placed in *A. (Actinocamax)*, and large species usually with a short cone-shaped alveolar fracture were assigned to *A. (Praeactinocamax)*. These subgenera were not recognized by Christensen (1982, 1986, 1991, 1993b, 1994). It is shown below, however, that the growth is isometric in the small species *A. (A.) verus*, whereas it is allometric in the large species *A. (P.) primus* Arkhangelsky and *A. (P.) plenus* from the Cenomanian, in addition to the medium-sized *A. (P.) cobbani* Christensen from the Coniacian (see Christensen 1974, 1990a, 1993b). In the present paper, therefore, two subgenera of *Actinocamax* are recognized. *A. (Actinocamax)* is characterized by its small size, usually long cone-shaped alveolar fracture, and isometric growth, and *A. (Praeactinocamax)* is distinguished by its medium to large size, usually short cone-shaped alveolar fracture, and allometric growth.

Distribution. – *Actinocamax* occurs from the Lower Cenomanian to middle Lower Campanian of the North European and North American Provinces.

Subgenus *Actinocamax (Actinocamax)* Miller, 1823

Emended diagnosis. – Small *Actinocamax* (mean value of the length of the guard 30–35 mm, and maximum length of guard 55 mm) with isometric growth; generally long, cone-shaped alveolar fracture; ventral fissure, furrow, and notch absent; single lateral furrows weakly developed or absent; granulation generally present.

Discussion. – *A. (Actinocamax)* has been the subject of excessive subdivision by Russian palaeontologists (Arkhangelsky 1912; Naidin 1952, 1953, 1964b; Nikitin 1958; Reyment & Naidin 1962; Glazunova 1972). Four species and ten subspecies have been established. *A. (A.) verus* from the Turonian to basal Lower Campanian includes six subspecies: the nominotypical subspecies, *dnestrensis* Naidin, 1952, *fragilis* Arkhangelsky, 1912, and *antefragilis* Naidin, 1964, which occur in NW Europe and the Russian Platform. The subspecies *subfragilis* Naidin, 1964, and *puschkariensis* Nikitin, 1958, are recorded only from the Russian Platform.

A. (A.) laevigatus Arkhangelsky, 1912, is from the basal Lower Campanian and occurs in the so-called 'Pteria' beds of the Russian Platform and possibly in western Europe (Naidin 1964b). It includes three subspecies: the nominotypical subspecies, *laevigatiformis* Naidin, 1964, and *pseudolaevigatus* Naidin, 1964.

Moreover, Naidin (1953) established *A. quasiverus* from the Upper Santonian of Crimea, and Glazunova (1972) erected *A. minutus* from the Lower Santonian of the Ulyanovsk district.

As shown below, many of these species and subspecies should be regarded as synonyms.

Distribution. – *Actinocamax (Actinocamax)* ranges from the Turonian to the middle Lower Campanian. It is distributed in the North European and North American Provinces.

Actinocamax (Actinocamax) verus verus Miller, 1823

Pl. 1:1–2

Synonymy. – ☐1823 *Actinocamax verus* – Miller, p. 64, Pl. 9:17–18 (*non* Pl. 3:16–20). ☐1876 *Actinocamax verus* Miller – Schlüter, p. 191, Pl. 52:9–15. ☐1885 *Actinocamax verus* Miller – Moberg, pp. 45–48, Pl. 4:15–26. ☐1897 *Actinocamax verus* Miller – Stolley, p. 292, Pl. 4:2–5. ☐1912 *Actinocamax verus* Miller var. *fragilis* n. var. – Arkhangelsky, p. 579, Pl. 9:15–17. ☐1921 *Actinocamax verus* Miller – Ravn, p. 37, Pl. 1:13–14. ☐1952 *Actinocamax verus* Miller – Naidin, p. 63, Pl. 1:7–8, 14. ☐1952 *Actinocamax verus* Miller var. *dnestrensis* n. var. – Naidin, p. 66, Pl. 1:9–10, Pl. 2:1–2. ☐1952 *Actinocamax verus* Miller var. *fragilis* Arkhangelsky – Naidin, p. 64, Pl. 1:11–12. ☐1957 *Actinocamax verus* Miller – Birkelund, p. 24, Pl. 2:4. ☐1958 *Actinocamax verus* Miller var. *puschkariensis* n. var. – Nikitin, p. 11, Pl. 2:13, 15. ☐1958 *Actinocamax verus* Miller var. *fragilis* Arkhangelsky – Nikitin, p. 9, Pl. 2:1–10,

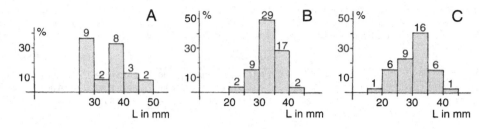

Fig. 17. Histograms of the length of the guard (L) of three samples of *Actinocamax verus verus* Miller. The figures above the bars are the number of specimens. □A. Bavnodde Greensand, Lower Santonian. □B. Eriksdal, upper Middle Santonian. □C. Kullemölla, basal Lower Campanian.

12. □1959 *Actinocamax verus* Miller – Naidin, p. 202, Pl. 19:11–12. □1962 *Actinocamax verus verus* Miller – Kongiel, p. 112. □1962 *Actinocamax verus dnestrensis* Naidin – Kongiel, p. 113, Pl. 20:14–17. □1962 *Actinocamax verus fragilis* Arkhangelsky – Reyment & Naidin, p. 153, figs 2.1–7, 10–11, 15–16. □1962 *Actinocamax verus verus* Miller – Reyment & Naidin, p. 155, figs 2.8–9, 17. □1964b *Actinocamax (Actinocamax) verus fragilis* Arkhangelsky – Naidin, p. 21, figs 5.2, 6–8, 11–12, 16–17. □1964b *Actinocamax (Actinocamax) verus verus* Miller – Naidin, p. 28, figs 5.9–10, 18. □1964b *Actinocamax (Actinocamax) verus dnestrensis* Naidin – Naidin, p. 29. □1972 *Actinocamax (Actinocamax)* Miller var. *fragilis* Arkhangelsky – Glazunova, p. 103, Pl. 42:3–7. □1972 *Actinocamax (Actinocamax) minutus* sp.nov. – Glazunova, p. 104, Pl. 42:2. □1973 *Actinocamax verus* Miller – Christensen, p. 133, Pl. 10:10. □1974 *Actinocamax (Actinocamax) verus verus* Miller – Naidin, p. 204, Pl. 72:2–3. □1974 *Actinocamax (Actinocamax) verus fragilis* Arkhangelsky – Naidin, p. 204, Pl. 72:4–6. □1974 *Actinocamax verus* Miller – Ernst & Schulz, p. 51. □1975a *Actinocamax verus* Miller – Christensen, p. 34, Pl. 2:1–2. □1979 *Actinocamax (Actinocamax) verus fragilis* Arkhangelsky – Naidin, p. 81, Pl. 3:1–4. □1986 *Actinocamax verus* Miller – Christensen, p. 23, Pl. 1:1. □1991 *Actinocamax verus* Miller – Christensen, p. 707, Pl. 1:1–9. □1993a *Actinocamax verus* Miller – Christensen, p. 44, Fig. 6A. □1994 *Actinocamax verus* Miller – Christensen, p. 152, Pl. 1:1–3.

Type. – Lectotype, by subsequent designation of Christensen (1991, p. 707), the original of Miller (1823, Pl. 9:17–18).

Material. – Twenty-eight specimens from the Bavnodde Greensand, including MGUH 23718 and GPIK 3901; 22 specimens are from west of Bavnodde (beds 1–6) and six specimens are from east of Bavnodde (beds 21–28).

Description. – See Christensen (1975a).

Biometry. – A sample consisting of 24 complete or near-complete specimens from west and east of Bavnodde was analyzed by univariate and bivariate methods.

Univariate analysis. – The results of the univariate analysis are given in Table 8. A histogram of the length of the

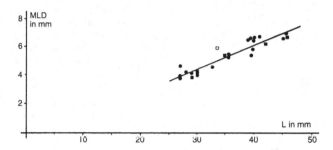

Fig. 18. Scatter plot and regresssion line of the length of the guard (L) *vs.* the maximum lateral diameter (MLD) of *Actinocamax verus verus* Miller from the Bavnodde Greensand, west of Bavnodde (dots) and east of Bavnodde (squares). One specimen of *A. verus* cf. *antefragilis* (open square) from the Arnager Limestone Formation at Stampe Å is also plotted.

guard is shown in Fig. 17. It is tested for normality using the Kolmogorov–Smirnov test for goodness of fit, and the test shows that the size-frequency distribution does not differ from normality at the 5% level ($D=0.2171$, with 24 degrees of freedom; $0.20>P>0.10$).

Bivariate analysis. – The scatter plot and regression line of the length of the guard *vs.* the maximum lateral diameter is shown in Fig. 18. The equation of the regression line is reported in Table 9. The correlation coefficient is very highly significant, $P<0.001$, with 22 degrees of freedom. The *t*-test on the *y*-intercept gives a value of 1.7232, with 22 degrees of freedom, which is not significant ($0.10>P>0.05$), implying an isometric relationship of the variates.

Discussion. – Christensen (unpublished) analyzed two samples of *A. verus* from Scania, southern Sweden: a sample consisting of 59 specimens from the Eriksdal marl, upper Middle Santonian, and a sample consisting of 39 specimens from the Kullemölla marl, basal Lower Campanian. The results of the univariate analyses are given in Tables 10–11.

Histograms of the length of the guard of these samples are shown in Fig. 17, and the equations of the regression lines of the length of the guard *vs.* the maximum lateral diameter are reported in Table 9. The relationship of the two variates is isometric in both samples.

Table 8. Univariate analysis of *Actinocamax v. verus* from the Bavnodde Greensand, east and west of Bavnodde. Measurements in millimetres.

Character	N	X̄	SD	CV	OR
L	24	35.5	6.3	17.7	27.0–45.5
MLD	24	5.3	1.1	21.3	3.8–7.0
L/MLD	24	6.7	0.5	7.5	5.8–7.6

Table 9. Estimates of statistical parameters of three regression analyses of the length of the guard (L) *vs.* the maximum lateral diameter (MLD) of *Actinocamax v. verus* from the Bavnodde Greensand (1), Eriksdal marl, Scania (2), and Kullemölla marl, Scania (3). Measurements in millimetres.

$y = a + bx$	N	r	Probability	SD_a	SD_b	SD_{yx}	t_a	Probability
1. MLD = −0.7286 + 0.1708L	24	0.9446	$P<0.001$	0.4228	0.0117	0.3539	1.7232	$0.10 > P > 0.05$
2. MLD = −0.2676 + 0.1490L	59	0.8092	$P<0.001$	0.4745	0.0143	0.4470	0.5640	$0.60 > P > 0.50$
3. MLD = −0.5441 + 0.1539L	39	0.8309	$P<0.001$	0.5455	0.0169	0.5304	0.9975	$0.40 > P > 0.30$

Table 10. Univariate analysis of *Actinocamax v. verus* from Kullemölla, Scania, basal Lower Campanian. Measurements in millimetres.

Character	N	X̄	SD	CV	OR
L	39	31.1	5.1	16.3	19.6–41.0
MLD	39	4.2	1.0	22.7	2.3–6.1
L/MLD	39	7.5	0.9	12.6	5.4–10.0

Table 11. Univariate analysis of *Actinocamax v. verus* from Eriksdal, Scania, upper Middle Santonian. Measurements in millimetres.

Character	N	X̄	SD	CV	OR
L	59	33.3	4.1	12.3	21.8–43.9
MLD	59	4.7	0.8	16.2	2.9–6.7
L/MLD	59	7.2	0.7	9.9	6.0–10.0

Table 12. Estimates of the relative abundance of three groups of *Actinocamax v. verus*. Group 1, specimens with a short cone-shaped alveolar fracture (*verus*-like specimens); group 2, specimens with a long cone-shaped alveolar fracture (*fragilis*-like specimens); group 3, specimens with a shallow pseudo-alveolus (*dnestrensis*-like specimens). Sources: 1, Christensen (1986); 2, Christensen (1991); 3, Christensen (1994); 4, this paper.

Locality	Age	Sediment	Source	Group 1	Group 2	Group 3	Σ
Kullemölla, Scania	basal Campanian	sandy marl	1	122 (79%)	26 (17%)	6 (4%)	154
Margate, Kent	Lower Upper Santonian	chalk	2	61 (77%)	16 (20%)	2 (3%)	79
Eriksdal, Scania	Upper Middle Santonian	sandy marl	1	100 (86%)	9 (8%)	7 (6%)	116
Red Lion Pit, Gravesend	basal Santonian	chalk	2	80 (94%)	1 (1%)	4 (5%)	85
Lonzée, Belgium	Santonian	greensand	3	22 (35%)	40 (63%)	1 (2%)	63
Bavnodde Greensand	Lower Santonian	greensand	4	5 (18%)	22 (78%)	1 (4%)	28

The means of the L/MLD ratio of the samples from Eriksdal and the Bavnodde Greensand were compared with the following result. The *F*-test gives a value of 1.9600, with 58 and 23 degrees of freedom, which is not significant, $0.10 > P > 0.05$. The *t*-test gives a value of 3.1801, with 81 degrees of freedom, which is highly significant, $0.005 > P > 0.001$. The specimens from the Bavnodde Greensand are thus more stout than the specimens from Eriksdal. The specimens from Kullemölla are more slender than those from Eriksdal. This suggests a trend through time towards more slender specimens. There may also be a trend towards smaller specimens through time (cf. the means and observed ranges of the length of the guard), although the differences are small. These suggestions should be tested by further analyses.

A. verus differs from *A. laevigatus* in having granules, which form corrugated transverse lines, and in having an alveolar fracture that is usually sharply demarcated from the surface of the guard. Specimens of *A. verus* with a flat anterior end or a shallow pseudoalveolus are distinguished from juvenile specimens of *Gonioteuthis westfalica* by being stouter and more lanceolate in ventral view.

A. v. verus has a short cone-shaped alveolar fracture, *A. verus fragilis* a long cone-shaped alveolar fracture, and *A. verus dnestrensis* a shallow pseudoalveolus. The variation of the structure of the alveolar end is due to differential calcification.

In previous papers, Christensen (1986, 1991, 1994) analyzed five samples of *A. verus* with respect to the structure of the alveolar end, and one sample from the Bavnodde Greensand is analyzed herein (Table 12). The following points are worthy of note. (1) Specimens with a shallow pseudoalveolus are very rare (2–6%) in all samples; (2) specimens with a short cone-shaped alveolar fracture predominate in the samples from marls of Scania and chalks of southern England (c. 80–95%), whereas specimens with a long cone-shaped alveolar fracture prevail in the samples from greensands of Belgium and Bornholm (c. 65–80%); and (3) the relative abundance of the specimens with a short or long cone-shaped alveolar fracture is not related to the age of the samples. The predominance of specimens with a long cone-shaped alveolar fracture in the samples from the greensands may be due to synsedimentary abrasion and/or diagenetic corrosion in these near-shore, shallow water deposits, which are rather coarse-grained and porous.

Christensen (1986, 1991, 1994) placed *A. verus dnestrensis* in synonymy with *A. v. verus*, because it was considered only as a variant. Christensen (1986, 1991) regarded *A. verus fragilis* to be a geographical subspecies, because Arkhangelsky (1912) and Naidin (1964b) noted that it prevails on the Russian Platform. However, Christensen (1994) placed *A. verus fragilis* in synonymy with *A. v. verus*, since specimens with a long cone-shaped alveolar fracture predominate in the Santonian greensand at

Lonzée near Gembloux in SE Belgium (Table 12). The majority of the specimens (more than three quarters) from the Bavnodde Greensand also have a long cone-shaped alveolar fracture. *A. verus fragilis* is consequently not to be considered as a geographical subspecies of the Russian Platform.

According to Nikitin (1958) *A. verus* var. *puschkariensis* var. nov. differs from the nominotypical subspecies by its larger guard and long cone-shaped alveolar fracture, which is pronouncedly asymmetric. The lectotype, here designated, is the original of Nikitin (1958, Pl. 2:13; No. 1952/1). This specimen is ca. 50 mm long and thus falls within the variation of *A. v. verus*. The two other characters are not diagnostic, and var. *puschkariensis* is therefore placed in synonymy with *A. v. verus*.

A. minutus differs in no significant respect from *A. v. verus* and is therefore placed in synonymy with this species. *A. quasiverus* was established by Naidin (1953) on the basis of one almost complete specimen and some fragments. It is distinguished from *A. verus* by its shape of the guard, absence of granules, and presence of striae. Naidin (1964b) noted that *A. quasiverus* is closely allied to specimens of *A. verus* from other parts of the Russian Platform.

Distribution. – *A. v. verus* is widespread in the North European Province. In NW Europe it is recorded mainly from the Santonian to the lower Lower Campanian. On Bornholm it enters in the upper Lower Santonian. In off-shore chalks it is most common in the Upper Santonian; it has an acme in the upper part of the *Uintacrinus socialis* Zone of southern England. On the Russian Platform it occurs from the Turonian to the lower Lower Campanian.

Actinocamax (Actinocamax) verus cf. *antefragilis* Naidin, 1964

Pl. 1:3

Synonymy. – ☐1897 *Actinocamax lundgreni* Stolley – Stolley, Pl. 3:15 (*non* Pl. 3:16–19 = *Goniocamax lundgreni*). ☐ cf. 1964b *Actinocamax (Actinocamax) verus antefragilis* subsp.nov. – Naidin, p. 28, Fig. 5.1. ☐1991 *Actinocamax verus* Miller – Christensen *in* Tröger & Christensen, p. 23.

Type. – Lectotype, here designated, 8023/3, the original of Naidin (1964b, Fig. 5.1), from the Lower Turonian, Surskoye, Ulyanovsk district.

Material. – GPIK 3902, glauconitic marl of the Arnager Limestone Formation, Stampe Å, loc. 5.

Description. – Guard almost complete but the most anterior and posterior part missing; length of guard estimated to 33.5 mm; guard club-shaped in ventral and lateral views with maximum diameters (5.9 mm) in posterior third; cross-section of guard at maximum lateral diameter circular; anterior end with an asymmetric, long cone-shaped alveolar fracture, the length of which is ca. one-third of length of the guard, and dorsal side more incised than ventral side; alveolar fracture merges gradually into the surface of the guard; dorsolateral double furrows well defined, extending from alveolar fracture and continuing to apical end; otherwise smooth guard.

Discussion. – GPIK 3902 is very similar to the specimen figured as *Actinocamax lundgreni* by Stolley (1897, Pl. 3:15), which came from the marl of the Arnager Limestone Formation at Muleby Å. Stolley (1897, p. 87) mentioned that this specimen is housed in the Geological Museum, Copenhagen, but it is not registered here. It is presumably lost.

According to Naidin (1964b) *A. verus antefragilis* is closely similar to *A. laevigatus* Arkhangelsky. Both species have a pronouncedly club-shaped guard and a long cone-shaped alveolar fracture, which merges gradually into the surface of the guard, and they are not granulated. GPIK 3902 is tentatively assigned to *A. verus antefragilis* owing to its shape, surface markings, and structure of the alveolar end, although it is stratigraphically younger. It is plotted in Fig. 18.

Distribution. – *A. verus* cf. *antefragilis* occurs in the marls of the Arnager Limestone Formation at Muleby Å and Stampe Å, which are upper Lower Coniacian and lower Middle Coniacian, respectively (see above). *A. verus antefragilis* is recorded from the Lower Turonian of Sursk and Ulyanovsk, Russian Platform.

Genus *Gonioteuthis* Bayle, 1878

Type species. – *Belemnites quadratus* Blainville, 1827, by original designation.

Diagnosis. – Medium-sized belemnitellids, with a shallow to deep pseudoalveolus (Riedel-Index 5–33; Riedel-Quotient 3–20); guard usually only slightly flattened or not flattened ventrally; generally subcylindrical or cylindrical in ventral view and cylindrical in lateral view; juvenile guard short and stout; dorsolateral longitudinal depressions and double furrows, vascular imprints, striae, and granules well developed; Schatzky distance small, usually 0.5–4.5 mm; bottom of ventral fissure commonly sine-shaped forming a large angle with the wall of the pseudoalveolus; vascular imprints branch off the dorsolateral double furrows posteriorly at an angle less than 30°.

Discussion. – The Middle Coniacian to Lower Campanian evolutionary lineage of *Gonioteuthis* includes, in ascending order, *G. praewestfalica* Ernst & Schulz, *G. w. westfalica* (Schlüter), *G.westfalicagranulata* (Stolley), *G. gran-*

ulata (Blainville), *G. granulataquadrata* (Stolley), *G. q. quadrata* (Blainville), and *G. quadrata gracilis* (Stolley). It has been studied in great detail by German authors, including Stolley (1897, 1916, 1930), Ernst (1964, 1966, 1968), Ernst & Schulz (1974), and Ulbrich (1971), in addition to Christensen (1975a, 1975b, 1986, 1988, 1991, 1994), Christensen & Schmid (1987), and Jarvis (1980).

The *Gonioteuthis* lineage provides a valuable tool for biostratigraphy. It is, however, necessary to analyse homogeneous samples of a certain size, say 10 specimens, in order to make a reliable specific determination, and a few specimens have only little stratigraphical value (Ernst 1964, Christensen 1975a).

Ernst (1963a, 1963b, 1964, 1966, 1968) and Ernst & Schulz (1974) characterized samples of *Gonioteuthis* on the basis of the mean values of various ratios, including the Riedel-Quotient, Riedel-Index, Slenderness-Quotient, and Flattening-Quotient (see above).

Christensen (1991) analyzed the growth relationship of the length of the guard *vs.* the depth of the pseudoalveolus, and the length of the guard *vs.* the dorsoventral diameter at the alveolar end, of a large number of samples, representing all species and subspecies of *Gonioteuthis* except *G. praewestfalica*. He showed that the relationship of the length of the guard *vs.* the depth of the pseudoalveolus generally may be regarded as isometric. The *Gonioteuthis* zonation of Ernst (1964), which is based on the mean Riedel-Quotient, is therefore valid. This zonation has been shown later to be useable by Ernst (1966, 1968), Ernst & Schulz (1974), Ulbrich (1971), Christensen (1975a, 1975b, 1986, 1988, 1991), and Jarvis (1980).

The relationship of the length of the guard *vs.* the dorsoventral diameter at the alveolar end is allometric to strongly allometric in most samples. It is therefore not valid to calculate the mean Slenderness-Quotient of samples which contain juvenile specimens.

The earliest species, *G. praewestfalica*, differs from younger species of the lineage by having a guard, which is markedly flattened ventrally and club-shaped in ventral view and generally does not possess granules.

Naidin (1964b) recognized two subgenera of *Gonioteuthis*: *G.* (*Gonioteuthis*), type species *Belemnites quadratus* Blainville, 1827, and *G.* (*Goniocamax*) Naidin, 1964, type species *Actinocamax lundgreni* Stolley, 1897. In addition to *lundgreni*, Naidin assigned early members of the *Gonioteuthis* lineage (*westfalica* and *westfalicagranulata*) as well as other species from the Russian Platform, such as *medwedicicus* Naidin, 1964, *matesovae* Naidin, 1964, and *intermedius* Arkhangelsky, 1912, to the subgenus *Goniocamax*. Moreover, he tentatively placed additional species from the North European and North American Provinces in the subgenus *Goniocamax*. The classification of Naidin was criticized by Ernst & Schulz (1974), who suggested that only the members of the evolutionary lineage mentioned above should be placed in the genus *Gonioteuthis*

and that the subgenus *Goniocamax* should be raised to the rank of a genus or considered a subgenus of *Belemnitella* d'Orbigny. They also suggested that only the *lundgreni* group and its ancestors should be assigned to *Goniocamax*. Christensen (1982, 1986, 1991) did not follow the classification of Naidin, because it is not logical from a phylogenetic point of view.

In the present paper, we follow the suggestion of Ernst & Schulz (1974) and only place the species of the evolutionary lineage mentioned above in *Gonioteuthis*. Moreover, *Goniocamax* is raised to the rank of a genus; it is discussed further below.

Ernst (1964) and Ernst & Schulz (1974) showed that one of the trends in the evolution of the *Gonioteuthis* lineage is the gradually increasing depth of the pseudoalveolus through time. As shown below, however, the depth of the pseudoalveolus of *G. praewestfalica* from the upper part of the Coniacian and lower Lower Santonian is larger than in *G. w. westfalica* from the lower part of the upper Lower Santonian and closely similar to *G. w. westfalica* from the upper part of the upper Lower Santonian. Moreover, it is shown below that the depth of the pseudoalveolus first decreases gradually, and that this trend is reversed in the lower part of the upper Lower Santonian.

Since *G. praewestfalica* has a deeper pseudoalveolus than *G. w. westfalica* from the lower and middle part of the upper Lower Santonian, it may have evolved from a species with a shallow pseudoalveolus, e.g., *Goniocamax lundgreni*. *G. praewestfalica* and *G. lundgreni* share certain morphological characters, i.e. the ventrally flattened and club-shaped guard. In addition, *G. praewestfalica* is very rarely granulated in contrast to later species of the *Gonioteuthis* lineage. Therefore, we tentatively suggest that *Gonioteuthis* is derived from *G. lundgreni*.

Distribution. – *Gonioteuthis* occurs from the upper Middle Coniacian to the boundary between the Lower and Upper Campanian. The genus had its evolutionary centre in northwestern Europe and is found almost exclusively in the Central European Subprovince. A few specimens are recorded from the northernmost part of the Tethyan Realm.

Gonioteuthis praewestfalica Ernst & Schulz, 1974

Pl. 1:4–5

Synonymy. – □1974 *Gonioteuthis westfalica praewestfalica* n. ssp. – Ernst & Schulz, p. 49, Pl. 5:2–9. □1983 *Gonioteuthis praewestfalica* Ernst & Schulz – Bailey *et al.*, p. 35. □1991 *Gonioteuthis westfalica praewestfalica* Ernst & Schulz – Christensen, p. 713, Pl. 1:10–13. □1991 *Gonioteuthis westfalica praewestfalica* Ernst & Schulz – Wood &

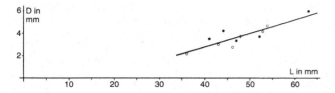

Fig. 19. Scatter plot and regression line of the length of the guard (L) *vs.* the depth of the pseudoalveolus (D) of *Gonioteuthis praewestfalica* from Jydegård (dots) and Horsemyre Odde (open circles). +=mean values.

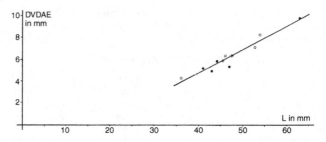

Fig. 20. Scatter plot and regression line of the length of the guard (L) *vs.* the dorsoventral diameter at the alveolar end (DVDAE) of *Gonioteuthis praewestfalica* from Jydegård (dots) and Horsemyre Odde (open circles). +=mean values.

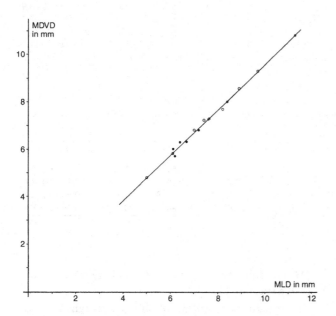

Fig. 21. Scatter plot and regression line of the maximum lateral diameter (MLD) *vs.* the maximum dorsoventral diameter (MDVD) of *Gonioteuthis praewestfalica* from Jydegård (dots) and Horsemyre Odde (open circles). +=mean values.

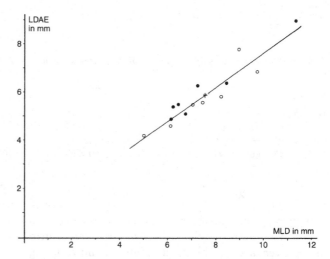

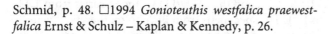

Fig. 22. Scatter plot and regression line of the maximum lateral diameter (MLD) *vs.* the lateral diameter at the alveolar end (LDAE) of *Gonioteuthis praewestfalica* from Jydegård (dots) and Horsemyre Odde (open circles). +=mean values.

Schmid, p. 48. □1994 *Gonioteuthis westfalica praewestfalica* Ernst & Schulz – Kaplan & Kennedy, p. 26.

Holotype. – By original designation, GPIH 1728, the original of Ernst & Schulz (1974, Pl. 5:4), from the Middle Coniacian *involutus/bucailli* Zone of Lägerdorf.

Material. – Eight specimens from Jydegård, including GPIK 3903, and seven specimens from Horsemyre Odde, including GPIK 3904; Bavnodde Greensand.

Dimensions. – See Table 13.

Description. – Guard small and flattened ventrally (Flattening-Quotient 1.02–1.09, mean value 1.05); club-shaped in ventral view, with the maximum lateral diameter in the middle part or posterior third, and slightly lanceolate in lateral view; anterior end flat, conical with a central shallow pseudoalveolus, or with a shallow pseudoalveolus (Riedel-Index ca. 6–10; mean value ca. 8; Riedel-Quotient ca. 10–16; mean value ca. 13); anterior

end with a ventral notch and sometimes with a dorsal notch; apex with indistinctly separated mucro, slightly displaced towards dorsal side; guard slender (Slenderness-Quotient 6.5–9); cross-section of anterior end subtriangular and compressed laterally; dorsolateral longitudinal depressions and double furrows, single lateral furrows, and longitudinal striae present; very few specimens with granules.

Biometry. – Fourteen specimens from Horsemyre Odde and Jydegård are analyzed by univariate and bivariate statistical methods.

Univariate analysis. – The results of the univariate analysis are shown in Table 14.

Bivariate analyses. – The following regression analyses are made: (1) the length of the guard *vs.* the depth of the pseudoalveolus, (2) the length of the guard *vs.* the dorsoventral diameter at the alveolar end, (3) the maximum lateral

Table 13. Dimensions of *Gonioteuthis praewestfalica* from the Bavnodde Greensand. GPIK 3903, Jydegård; GPIK 3904, Horsemyre Odde. Measurements in millimetres.

Character	GPIK 3903	GPIK 3904
L	47.0	46.0
D	3.3	2.8
DVDAE	5.3	6.3
LDAE	5.1	5.6
MLD	6.7	7.4
MDVD	6.3	7.2
RQ	14.2	16.4
RI	7.0	6.1
SQ	8.9	7.3
L/MLD	7.0	6.2
FQ	1.06	1.03
MLD/LDAE	1.31	1.32
DVDAE/LDAE	1.04	1.13

Table 14. Univariate analysis of *Gonioteuthis praewestfalica* from the Bavnodde Greensand. Measurements in millimetres.

Character	N	$\bar{X}$	SD	CV	OR
L	10	47.8	7.7	16.1	36.0–63.0
D	14	3.9	1.2	30.3	2.2–6.0
DVDAE	13	6.2	1.5	23.8	4.2–9.7
LDAE	14	5.9	1.3	21.7	4.2–9.0
MLD	14	7.5	1.7	22.5	5.0–11.3
MDVD	14	7.2	1.6	22.5	4.8–10.8
RQ	10	13.2	2.2	16.7	10.5–16.4
RI	10	7.8	1.3	16.6	6.1–9.6
SQ	9	not calculated			6.5–8.9
L/MLD	10	not calculated			5.4–7.2
FQ	14	1.05	0.02	1.8	1.02–1.09
MLD/LDAE	14	1.26	0.09	7.5	1.14–1.41
DVDAE/LDAE	13	1.05	0.04	3.8	0.98–1.13

Table 15. Estimates of statistical parameters of four regression analyses of *Gonioteuthis praewestfalica* from the Bavnodde Greensand. Measurements in millimetres.

$y=a+bx$	N	r	Probability	SD_a	SD_b	SD_{yx}	t_a	Probability
D=−2.1183+0.1229L	10	0.8781	$P<0.001$	1.1340	0.0234	0.5427	1.8679	$0.10>P>0.05$
DVDAE=−3.6678+0.2104L	9	0.9639	$P<0.001$	1.0448	0.0218	0.4952	3.5106	$0.01>P>0.005$
MDVD=−0.0476+0.9614MLD	13	0.9975	$P<0.001$	0.2225	0.0287	0.1688	0.2139	$0.90>P>0.80$
LDAE=0.5523+0.7199MLD	14	0.9390	$P<0.001$	0.5587	0.0731	0.4421	0.9886	$0.40>P>0.30$

diameter *vs.* the maximum dorsoventral diameter, and (4) the maximum lateral diameter *vs.* the lateral diameter at the alveolar end. The scatter plots are shown in Figs. 19–22, as are the regression lines, and the equations of the regressions lines are given in Table 15.

The correlation coefficients are very highly significant in all analyses ($P<0.001$, with $N-2$ degrees of freedom) (Table 15). The relationships of the length of the guard *vs.* the depth of the pseudoalveolus, the maximum lateral diameter *vs.* the maximum dorsoventral diameter, and the maximum lateral diameter *vs.* the lateral diameter at the alveolar end, may be regarded as isometric, whereas the relationship of the length of the guard *vs.* the dorsoventral diameter at the alveolar end is highly allometric (Table 15).

Discussion. – *G. praewestfalica* was fully described by Ernst & Schulz (1974) on the basis of German material. They considered *praewestfalica* as a subspecies of *G. westfalica* and noted that *praewestfalica* can be distinguished from *westfalica* only on the basis of a biometric analysis of a sample. The main characters separating *praewestfalica* from *westfalica* are the ventrally flattened guard, which is club-shaped in ventral view. Moreover, *G. praewestfalica* is very rarely granulated, and less than 10% of the specimens of samples of *G. praewestfalica* from Lägerdorf and Bornholm are granulated (Schulz, unpublished; herein). In contrast, about the half of the specimens of populations of *G. westfalica* are granulated (Ernst & Schulz

1974). The subspecies *praewestfalica* is here raised to the rank of a species.

The specimens from Bornholm are assigned to *G. praewestfalica*, because the guard is markedly flattened ventrally and club-shaped in ventral view. One specimen from Bornholm is granulated. The Flattening-Quotient of the German specimens varies from 1.00 to 1.04, and the MLD/LDAE Index varies from 1.10 from 1.35. The Flattening-Quotient of the Bornholm specimens varies from 1.02 to 1.09, and the MLD/LDAE Index varies from 1.14 to 1.41. The specimens from Bornholm, therefore, generally are more flattened ventrally and more lanceolate in ventral view than the specimens from NW Germany. Nevertheless, we assign the specimens from Bornholm to *G. praewestfalica*. The mean Riedel-Index of the sample of *G. praewestfalica* from Bornholm is 7.8 (mean Riedel-Quotient ca. 13), which is very closely similar to that of *G. praewestfalica* from Lägerdorf (mean Riedel-Index is ca. 7.9, and mean Riedel-Quotient is ca. 13; see Table 30 and Ernst & Schulz 1974). *G. praewestfalica* is discussed further below.

Distribution. – *G. praewestfalica* is recorded from Lägerdorf (Ernst & Schulz 1974), Helgoland (Wood & Schmid 1991), the Münster Basin (Kaplan & Kennedy 1994), Kent, southern England (Bailey *et al.* 1983; Christensen 1991), and Bornholm (this paper). It occurs in the upper Middle and Upper Coniacian of NW Germany and England. On Bornholm, it occurs together with *Goniocamax l. lundgreni*, *G. birkelundae* sp.nov., *Belemnitella schmidi* sp.nov., and *B. p. propinqua* in beds from the lower Lower Santonian (Fig. 12). *G. praewestfalica* thus appears later on Bornholm than at Lägerdorf.

Gonioteuthis westfalica westfalica (Schlüter, 1874)

Pl. 1:6–11

Synonymy. – See Christensen (1975a).

Type. – Lectotype, by subsequent designation of Ernst & Schulz (1974, p. 50), the original of Schlüter (1876, Pl. 53:10), from the 'Emschermergel der Zeche Blücher bei Horst', north of Essen, Westfalia.

Material. – West of Forchhammers Odde, 35 specimens; Forchhammers Odde, 31 specimens; east of Bavnodde, 97 specimens, including MGUH 23719-20; and west of Bavnodde, several hundreds of specimens, including MGUH 23721-4.

Description. – Guard small and usually not flattened ventrally (Flattening-Quotient 0.92–1.08, mean values 0.99–1.01); generally subcylindrical or slightly lanceolate in ventral view and cylindrical in lateral view; guard slender (Slenderness-Quotient 6–10); shape variable because of allometric growth, adult specimens more stout and more lanceolate in ventral view than juvenile specimens; anterior end flat, cone-shaped with a central, very shallow pseudoalveolus, or with a very shallow pseudoalveolus (Riedel-Index generally 4–10, mean values 6.5–8; Riedel-Quotient generally 8–20, mean values 14–17); anterior end with a ventral notch and/or furrow and sometimes with a dorsal notch; apex with indistinctly separated mucro, slightly displaced towards dorsal side; cross-section of anterior end pointed oval or subtriangular and compressed laterally; dorsolateral longitudinal depressions and double furrows, single lateral furrows, and longitudinal striae present; half the specimens with granules; Schatzky distance small, usually 1.5–2.5 mm; and bottom of ventral fissure commmonly sine-shaped forming a large angle, often more 90°, with wall of pseudoalveolus.

Biometry. – Six samples from the Bavnodde Greensand are analyzed by univariate and bivariate statistical methods: (1) 34 specimens from west of Forchhammers Odde; (2) 26 specimens from Forchhammers Odde; (3) 97 specimens from east of Bavnodde; (4) 151 specimens from west of Bavnodde, between beds 4 and 7; (5) 188 specimens from west of Bavnodde, between beds 2 and 4; and (6) 70 specimens from west of Bavnodde, between beds 1 and 2.

The specimens from east of Bavnodde were subdivided into two groups: one group from below bed 24 and one group from above bed 24 (Fig. 6). The means of the Riedel-Index of the two groups were compared by the *t*-test, which showed that they do not differ significantly at the 5% level. The specimens are therefore lumped in the statistical analyses.

Univariate analyses. – The results of the univariate analyses are given in Tables 16–21. Histograms of the length of the guard, Riedel-Quotient, Riedel-Index, MLD/MDVD Index, and MLD/LDAE Index of the four samples from east and west of Bavnodde are shown in Figs. 23–27.

Bivariate analyses. – The following regression analyses are made: (1) the length of the guard *vs.* the depth of the pseudoalveolus, (2) the length of the guard *vs.* the dorsoventral diameter at the alveolar end, (3) the maximum lateral diameter *vs.* maximum dorsoventral diameter, (4) and the maximum lateral diameter *vs.* lateral diameter at the alveolar end. The equations of the regression lines are given in Tables 22–27, and scatter plots are shown in Figs. 28–43.

The correlation coefficients are tested for significance, and the *y*-intercepts are tested by *t*-tests to see if they differ significantly from zero (Tables 22–27).

The coefficients of determination, r^2, are generally very low (0–10%) in the analyses of the length of the guard *vs.* the depth of the pseudoalveolus. The sample from west of Forchhammers Odde differs from the other samples, since the coefficient of determination is 43%. The coefficients of determination are larger in the analyses of the length of the guard *vs.* the dorsoventral diameter at the alveolar end (70–95%), and the maximum lateral diameter *vs.* the lateral diameter at the alveolar end (85–95%). They are very large in the analyses of the maximum lateral diameter *vs.* the maximum dorsoventral diameter (98–99.5%).

Length of guard vs. depth of pseudoalveolus. – The correlation coefficients are not significant in three out of six samples. The correlation is not significant in the samples from Forchhammers Odde, west of Bavnodde, between beds 2 and 4, and west of Bavnodde, between beds 1 and 2. Regression analyses are not performed in these samples. In the samples from east of Bavnodde and west of Bavnodde, between beds 4 and 7, the values of the correlation coefficient are low but the correlation coefficients are significant. In the samples from west of Forchhammers Odde and east of Bavnodde, the relationship of the variates may be regarded as isometric, whereas it is strongly allometric in the sample from west of Bavnodde, between beds 4 and 7.

Christensen (1975a, p. 38) analyzed the relationship of the length of the guard *vs.* the depth of the pseudoalveolus of a large sample (*N*=216) of *G. w. westfalica* from the lower '*westfalica* beds' from Essen-Vogelheim (Ernst 1964, p. 118). In this sample, the value of the correlation

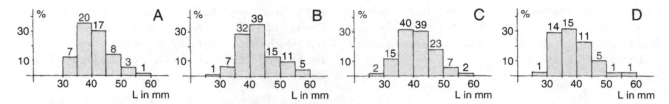

Fig. 23. Histograms of the length of the guard (L) of four samples of *Gonioteuthis w. westfalica* from east and west of Bavnodde. □A. East of Bavnodde. □B. West of Bavnodde, between beds 4 and 7. □C. West of Bavnodde, between beds 2 and 4. □D. West of Bavnodde, between beds 1 and 2. The figures above the bars are the actual number of specimens.

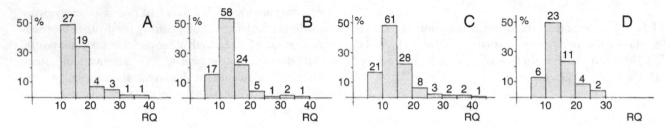

Fig. 24. Histograms of the Riedel-Quotient (RQ) of four samples of *Gonioteuthis w. westfalica* from east and west of Bavnodde. For explanation see Fig. 23.

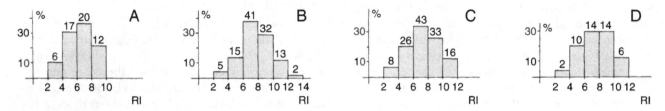

Fig. 25. Histograms of the Riedel-Index (RI) of four samples of *Gonioteuthis w. westfalica* from east and west of Bavnodde. For explanation see Fig. 23.

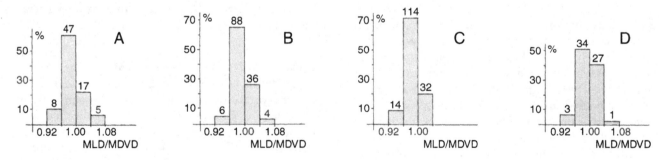

Fig. 26. Histograms of the Flattening-Quotient (MLD/MDVD) of four samples of *Gonioteuthis w. westfalica* from west and east of Bavnodde. For explanation see Fig. 23.

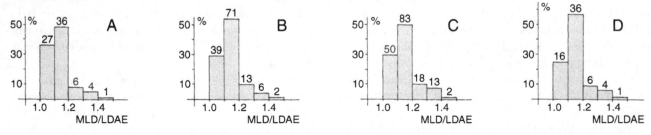

Fig. 27. Histograms of the MLD/LDAE Index of four samples of *Gonioteuthis w. westfalica* from east and west of Bavnodde. For explanation see Fig. 23.

Table 16. Univariate analysis of *Gonioteuthis w. westfalica* from the Bavnodde Greensand, west of Forchhammers Odde. Measurements in millimetres.

Character	N	X̄	SD	CV	OR
L	15	43.0	5.2	12.2	36.5−54.5
D	34	3.0	0.8	26.1	1.0−4.7
DVDAE	30	5.4	1.3	23.3	3.7−8.7
LDAE	30	5.0	1.1	21.9	3.5−8.2
MLD	28	5.6	1.4	25.1	3.8−8.6
MDVD	28	5.5	1.4	24.6	3.8−8.3
RQ	15	15.4	6.6	42.8	10.6−37.0
RI	15	7.2	1.7	24.1	2.7−9.4
SQ	14	not calculated			6.2−10.0
L/MLD	15	not calculated			5.8−9.7
FQ	29	1.01	0.02	2.4	0.98−1.06
MLD/LDAE	25	1.13	0.08	7.2	1.01−1.34
DVDAE/LDAE	30	1.07	0.04	3.8	1.00−1.17

Table 17. Univariate analysis of *Gonioteuthis w. westfalica* from the Bavnodde Greensand, Forchhammers Odde. Measurements in millimetres.

Character	N	X̄	SD	CV	OR
L	15	42.5	4.1	9.7	34.7−52.1
D	25	2.8	0.8	29.7	1.5−4.8
DVDAE	26	5.4	1.2	22.5	3.7−7.5
LDAE	26	5.1	1.0	20.4	3.7−7.0
MLD	24	5.7	1.3	22.1	3.9−8.5
MDVD	24	5.7	1.2	21.8	3.8−8.1
RQ	15	15.5	2.8	17.7	10.0−20.0
RI	15	6.7	1.3	20.1	5.0−10.0
SQ	15	not calculated			6.4−9.6
L/MLD	14	not calculated			6.2−8.5
FQ	24	1.01	0.02	1.9	0.98−1.05
MLD/LDAE	23	not calculated			1.05−1.29
DVDAE/LDAE	25	1.05	0.04	3.9	0.98−1.14

Table 18. Univariate analysis of *Gonioteuthis w. westfalica* from the Bavnodde Greensand, east of Bavnodde. Measurements in millimetres.

Character	N	X̄	SD	CV	OR
L	56	41.3	5.8	14.0	30.1−56.0
D	93	2.7	0.7	25.2	1.0−4.0
DVDAE	96	5.1	1.1	22.3	3.2−8.8
LDAE	97	4.8	1.0	19.8	3.3−7.7
MLD	77	5.5	1.1	20.8	3.5−8.7
MDVD	77	5.5	1.1	20.9	3.5−8.8
RQ	55	16.7	5.7	33.2	10.3−39.0
RI	55	6.5	1.8	26.9	2.6−9.8
SQ	54	not calculated			6.1−10.0
L/MLD	54	not calculated			5.9−9.3
FQ	77	1.00	0.03	2.5	0.95−1.07
MLD/LDAE	74	not calculated			1.02−1.49
DVDAE/LDAE	96	1.07	0.05	4.9	0.94−1.18

Table 19. Univariate analysis of *Gonioteuthis w. westfalica* from the Bavnodde Greensand, west of Bavnodde, between beds 4 and 7. Measurements in millimetres.

Character	N	X̄	SD	CV	OR
L	110	42.8	6.1	14.3	29.7−58.7
D	107	3.3	0.8	25.4	1.3−6.6
DVDAE	151	5.4	1.3	24.0	3.2−9.2
LDAE	151	5.0	1.2	22.9	3.2−8.8
MLD	135	5.8	1.3	23.1	3.5−9.4
MDVD	135	5.7	1.3	22.9	3.5−9.0
RQ	108	14.0	4.8	34.1	7.9−35.3
RI	108	7.8	2.1	27.1	2.8−12.7
SQ	109	not calculated			6.3−9.6
L/MLD	110	not calculated			6.0−9.3
FQ	134	1.00	0.02	2.1	0.95−1.08
MLD/LDAE	131	not calculated			1.00−1.43
DVDAE/LDAE	151	1.07	0.05	4.3	0.95−1.20

Table 20. Univariate analysis of *Gonioteuthis w. westfalica* from the Bavnodde Greensand, west of Bavnodde, between beds 2 and 4. Measurements in millimetres.

Character	N	X̄	SD	CV	OR
L	128	41.6	5.8	14.0	27.7−59.7
D	187	3.0	0.9	30.8	1.0−7.5
DVDAE	188	5.1	1.1	22.3	3.3−9.1
LDAE	188	4.8	1.0	20.4	3.2−8.3
MLD	160	5.6	1.3	22.3	3.4−9.4
MDVD	161	5.7	1.2	21.8	3.4−9.4
RQ	126	15.1	6.0	39.5	8.3−43.6
RI	126	7.4	2.1	28.9	2.3−12.0
SQ	125	not calculated			6.0−10.4
L/MLD	125	not calculated			5.5−10.9
FQ	160	0.99	0.02	2.0	0.92−1.04
MLD/LDAE	166	not calculated			1.00−1.50

Table 21. Univariate analysis of *Gonioteuthis w. westfalica* from the Bavnodde Greensand, west of Bavnodde, between beds 1 and 2. Measurements in millimetres.

Character	N	X̄	SD	CV	OR
L	48	39.0	6.3	16.2	27.6−58.5
D	65	2.8	0.8	28.4	1.4−5.2
DVDAE	69	4.9	1.1	22.6	3.3−9.0
LDAE	70	4.5	0.9	20.5	3.1−8.0
MLD	66	5.3	1.2	22.6	3.6−9.5
MDVD	66	5.3	1.2	22.5	3.5−9.7
RQ	46	14.3	4.5	31.1	8.4−25.9
RI	46	7.6	2.1	28.0	3.9−12.0
SQ	45	not calculated			6.0−9.8
L/MLD	47	not calculated			5.8−9.2
FQ	66	1.00	0.02	2.4	0.93−1.05
MLD/LDAE	63	not calculated			1.04−1.43

coefficient is low but highly significant ($r = 0.2826$; $0.01 > P > 0.001$; 194 degrees of freedom), and the relationship of the variates is isometric.

It thus seems that the values of the correlation coefficients are low and the correlation coefficients are not significant or significant in *G. w. westfalica*.

The means of the Riedel-Quotient and Riedel-Index were calculated in all samples, although it is not legal to do this in the sample from west of Bavnodde, between beds 4 and 7, because of the allometry of the variates. However, the coefficient of determination is only 4%.

Length of guard vs. dorsoventral diameter at alveolar end. – The correlation coefficients are very highly significant, and the relationship of the variates is allometric to very strongly allometric. Because of the allometry, juvenile specimens are more slender than adult specimens. For instance, if the correlation coefficient is perfect ($r = 1.0000$) in the sample from west of Bavnodde, between beds 4 and 7, the Slenderness-Quotient will vary

as shown in Table 28. Owing to individual variation, the Slendernes-Quotient varies from 6.3 (an adult specimen) to 9.6 (a juvenile specimen). The means of the Slendernes-Quotient, therefore, have not been calculated.

Maximum lateral diameter vs. maximum dorsoventral diameter. – The correlation coefficients are very highly significant, and the y-intercepts do not differ significantly from zero (=isometry). The slopes are tested by t-tests to see if they differ significantly from 1. The tests show that none of the slopes differ significantly from 1 at the 5% level. Since the y-intercepts do not differ significantly from zero and the slopes do not differ significantly from 1, the equation of the regression lines can be written $y = x$.

Maximum lateral diameter vs. lateral diameter at alveolar end. – The correlation coefficients are very highly significant, and the relationship of the variates is either isometric or allometric to very strongly allometric. It is isometric in the sample from west of Forchhammers Odde, allo-

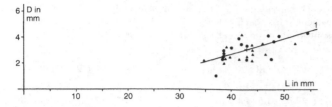

Fig. 28. Scatter plot and regression line of the length of the guard (L) *vs.* the depth of the pseudoalveolus (D) of *Gonioteuthis w. westfalica* from west of Forchhammers Odde (dots, regression line 1) and Forchhammers Odde (triangles). + = mean values.

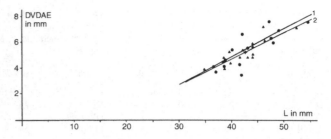

Fig. 29. Scatter plot and regression lines of the length of the guard (L) *vs.* the dorsoventral diameter at the alveolar end (DVDAE) of *Gonioteuthis w. westfalica* from west of Forchhammers Odde (dots, regression line 1) and Forchhammers Odde (triangles, regression line 2). + = mean values.

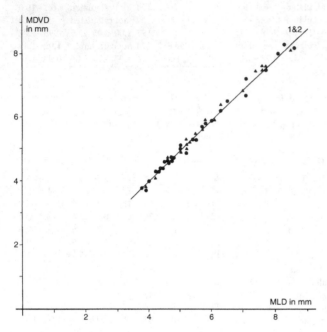

Fig. 30. Scatter plot and regression lines of the maximum lateral diameter (MLD) *vs.* the maximum dorsoventral diameter (MDVD) of *Gonioteuthis w. westfalica* from west of Forchhammers Odde (dots, regression line 1) and Forchhammers Odde (triangles, regression line 2). The two regression lines are virtually identical.

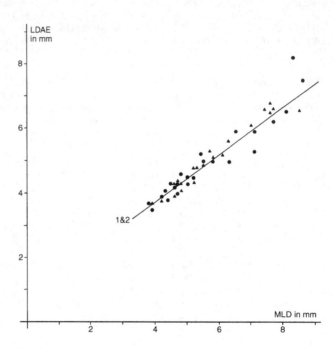

Fig. 31. Scatter plot and regression lines of the maximum lateral diameter (MLD) *vs.* the lateral diameter at the alveolar end (LDAE) of *Gonioteuthis w. westfalica* from west of Forchhammers Odde (dots, regression line 1) and Forchhammers Odde (triangles, regression line 2). The two regression lines are virtually identical.

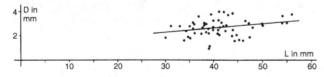

Fig. 32. Scatter plot and regression line of the length of the guard (L) *vs.* the depth of the pseudoalveolus (D) of *Gonioteuthis w. westfalica* from east of Bavnodde. + = mean values.

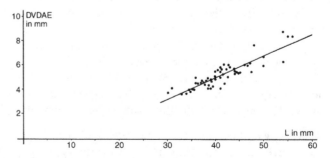

Fig. 33. Scatter plot and regression line of the length of the guard (L) *vs.* the dorsoventral diameter at the alveolar end (DVDAE) of *Gonioteuthis w. westfalica* from east of Bavnodde. + = mean value.

Table 22. Estimates of statistical parameters of four regression analyses of *Gonioteuthis w. westfalica* from the Bavnodde Greensand, west of Forchhammers Odde. Measurements in millimetres.

$y = a + bx$	N	r	Probability	SD_a	SD_b	SD_{yx}	t_a	Probability
D = -1.7483 + 0.1128L	15	0.6556	0.01 > P > 0.001	1.5646	0.0362	0.7064	1.1174	0.30 > P > 0.25
DVDAE = -3.8397 + 0.2176L	14	0.8805	P < 0.001	1.4997	0.0346	0.6767	2.5603	0.025 > P > 0.02
MDVD = 0.1317 + 0.9620MLD	28	0.9952	P < 0.001	0.1267	0.0221	0.1608	1.0392	0.40 > P > 0.30
LDAE = 0.7929 + 0.7341MLD	23	0.9221	P < 0.001	0.3822	0.0651	0.4532	2.0747	0.10 > P > 0.05

Table 23. Estimates of statistical parameters of four regression analyses of *Gonioteuthis w. westfalica* from the Bavnodde Greensand, Forchhammers Odde. Measurements in millimetres.

$y = a + bx$	N	r	Probability	SD_a	SD_b	SD_{yx}	t_a	Probability
D = -0.2820 + 0.0732L	15	0.4748	0.10 > P > 0.05	(no significant correlation)				
DVDAE = -3.3153 + 0.2023L	15	0.8489	P < 0.001	1.4738	0.0345	0.5341	2.2495	0.05 > P > 0.025
MDVD = 0.1056 + 0.9711MLD	24	0.9957	P < 0.001	0.1589	0.0271	0.1643	0.6646	0.60 > P > 0.50
LDAE = 0.8200 + 0.7383MLD	23	0.9794	P < 0.001	0.2943	0.0499	0.3002	2.7867	0.02 > P > 0.01

Table 24. Estimates of statistical parameters of four regression analyses of *Gonioteuthis w. westfalica* from the Bavnodde Greensand, east of Bavnodde. Measurements in millimetres.

$y = a + bx$	N	r	Probability	SD_a	SD_b	SD_{yx}	t_a	Probability
D = 1.2334 + 0.0348L	55	0.2828	0.05 > P > 0.02	0.6806	0.0163	0.7006	1.8123	0.10 > P > 0.05
DVDAE = -2.1199 + 0.1778L	54	0.8889	P < 0.001	0.5517	0.0132	0.5628	3.8424	P < 0.001
MDVD = 0.0174 + 0.9986MLD	77	0.9922	P < 0.001	0.0400	0.0072	0.0710	0.4350	0.70 > P > 0.60
LDAE = 0.5294 + 0.7814MLD	74	0.9346	P < 0.001	0.1797	0.0323	0.3183	2.9467	0.005 > P > 0.001

Table 25. Estimates of statistical parameters of four regression analyses of *Gonioteuthis w. westfalica* from the Bavnodde Greensand, west of Bavnodde, between beds 4 and 7. Measurements in millimetres.

$y = a + bx$	N	r	Probability	SD_a	SD_b	SD_{yx}	t_a	Probability
D = 2.0436 + 0.0290L	108	0.1949	0.05 > P > 0.02	0.6069	0.0141	0.8623	3.3671	0.005 > P > 0.001
DVDAE = -2.9262 + 0.1971L	109	0.9417	P < 0.001	0.2736	0.0063	0.4064	10.6966	P < 0.001
MDVD = 0.0955 + 0.9799MLD	134	0.9959	P < 0.001	0.0815	0.0138	0.2127	1.1713	0.25 > P > 0.20
LDAE = 0.4093 + 0.8053MLD	131	0.9558	P < 0.001	0.1117	0.0189	0.2899	3.6635	P < 0.001

Table 26. Estimates of statistical parameters of four regression analyses of *Gonioteuthis w. westfalica* from the Bavnodde Greensand, west of Bavnodde, between beds 2 and 4. Measurements in millimetres.

$y = a + bx$	N	r	Probability	SD_a	SD_b	SD_{yx}	t_a	Probability
D = 2.5833 + 0.0105L	127	0.0704	P > 0.10	(no significant correlation)				
DVDAE = -2.2404 + 0.1827L	127	0.8883	P < 0.001	0.3605	0.0086	0.5552	6.2139	P < 0.001
MDVD = 0.1829 + 0.9757MLD	83	0.9969	P < 0.001	0.1882	0.0335	0.3967	0.9718	0.40 > P > 0.30
LDAE = 0.8491 + 0.7118MLD	157	0.9324	P < 0.001	0.1394	0.0242	0.3826	6.0891	P < 0.001

Table 27. Estimates of statistical parameters of four regression analyses of *Gonioteuthis w. westfalica* from the Bavnodde Greensand, west of Bavnodde, between beds 1 and 2. Measurements in millimetres.

$y = a + bx$	N	r	Probability	SD_a	SD_b	SD_{yx}	t_a	Probability
D = 1.9441 + 0.0252L	46	0.2013	P > 0.10	(no significant correlation)				
DVDAE = -1.9855 + 0.1793L	45	0.9123	P < 0.001	0.4988	0.0127	0.5399	3.9807	P < 0.001
MDVD = 0.0563 + 0.9875MLD	66	0.9941	P < 0.001	0.0956	0.0176	0.1699	0.5887	0.60 > P > 0.50
LDAE = 0.6474 + 0.7400MLD	63	0.9535	P < 0.001	0.1565	0.0290	0.2756	4.1362	P < 0.001

metric in the sample from Forchhammers Odde, strongly allometric in the sample from east of Bavnodde, and very strongly allometric in the sample from west of Bavnodde. In the samples showing an allometric relationship of the variates, adult specimens generally are more lanceolate in ventral view than juvenile specimens. For instance, if the correlation coefficient is perfect ($r = 1.0000$) in the sample from west of Bavnodde, between beds 1 and 2, the ratio of the MLD/LDAE Index will vary as shown in Table 29. Because of individual variation, the MLD/LDAE Index

Table 28. Tabulation of the Slenderness-Quotient (SQ) with increasing length of the guard (L) in *Gonioteuthis w. westfalica* from the Bavnodde Greensand, west of Bavnodde, between beds 4 and 7. Owing to allometry, the Slenderness-Quotient will decrease when the length of the guard increases. See text for discussion.

L (in mm)	SQ
30	10.4
40	8.1
50	7.2
60	6.7

Table 29. Tabulation of the MLD/LDAE Index with increasing maximum lateral diameter in *Gonioteuthis w. westfalica* from the Bavnodde Greensand, west of Bavnodde, between beds 1 and 2. Owing to allometry, the MLD/LDAE Index will increase when the maximum lateral diameter increases. See text for discussion.

MLD (in mm)	MLD/LDAE
4	1.08
5	1.15
6	1.18
7	1.20
8	1.22
9	1.23

varies from 1.04 (an adult specimen) to 1.43 (an adult specimen). The means of the MLD/LDAE Index are consequently not calculated in the samples showing an allometric relationship of the variates.

Discussion. – The means of the length of the guard of the six samples are very closely similar (weighted grand mean of the six samples is ca. 42 mm), as is the maximum length of the guard, which is slightly less than 60 mm. Ernst & Schulz (1974) showed that the maximum length of the guard of most specimens of *G. w. westfalica* from the lower *westfalica* beds of Lägerdorf is 60 mm. On the other hand, Ernst (1964) demonstrated that in a large sample, consisting of 216 specimens of *G. w. westfalica* from Essen-Vogelheim in the Münster Basin, more than 25% of the specimens fall in the size range 60–70 mm.

Ernst & Schulz (1974, Fig. 15) showed that the mean Riedel-Index increases regularly from the upper Coniacian *G. praewestfalica* to the upper Middle Santonian *G. westfalicagranulata*. It is ca. 8 in *G. praewestfalica* and *G. w. westfalica* from the lower *westfalica* Zone, and ca. 8.5 in *G. w. westfalica* from the boundary between the lower and upper *westfalica* Zones. In the upper *westfalica* Zone it is ca. 9 in the lower part, ca. 9.5 in the middle part, and ca. 10.5 in the upper part of the zone. It is ca. 12 in *G. westfalicagranulata* from the middle part of the *westfalicagranulata* Zone. This trend continues stratigraphically upwards to the Lower Campanian *G. q. quadrata* as shown by Ernst (1964). In this context, it is noteworthy that *Gonioteuthis* is virtually absent in the lower Lower Santonian *pachti/undulatoplicatus* Zone of Lägerdorf (see above and Fig. 11).

The mean Riedel-Indices of the samples of *G. praewestfalica* and *G. w. westfalica* from the Lower Santonian belemnite assemblage zones 4–8 of Bornholm are shown in Table 30. Since the number of specimens is small in the samples from west of Forchhammers Odde and Forchhammers Odde, i.e. belemnite assemblage zone 6, we have lumped the specimens and analyzed them biometrically with respect to the Riedel-Index. The mean Riedel-Index is 6.9, the standard deviation is 1.5, and $N=30$. The t-test of the means of the Riedel-Index of the two samples showed that they do not differ significantly at the 5% level.

The mean Riedel-Index decreases regularly from *G. praewestfalica* in zones 4–5 (mean Riedel-Index 7.8) to *G. w. westfalica* in zone 7 (mean Riedel-Index 6.5), whereas it increases to 7.8 in the oldest sample of *G. w. westfalica* in zone 8. The weighted grand mean of the three samples from zone 8 is 7.6. Belemnite assemblage zones 4–6 of Bornholm equate with the *pachti/undulatoplicatus* Zone of Lägerdorf (Fig. 11).

G. praewestfalica thus has a deeper pseudoalveolus than *G. w. westfalica* from zones 6 and 7. On the other hand, the depth of the pseudoalveolus is closely similar in *G. praewestfalica* and *G. w. westfalica* from zone 8 (Table 30).

The means of the Riedel-Index of the samples of *G. praewestfalica* from zones 4–5 and *G. w. westfalica* from zone 6 were compared statistically, and they do not differ significantly at the 5% level. The means of *G. w. westfalica* from zones 6 and 7 also do not differ significantly at the 5% level.

Table 30. Means and observed ranges of the Riedel-Index of samples of *Gonioteuthis praewestfalica* and *G. w. westfalica* from Bornholm and Lägerdorf, NW Germany. The sample of *G. praewestfalica* from Lägerdorf includes the material of Ernst & Schulz (1974, Fig. 15), except a badly preserved specimen with a Riedel-Index of 1.9(?). In addition, one specimen with a Riedel-Index of 6.7 collected in 1974, and one specimen with a Riedel-Index of 11.2 collected in 1985 are included.

Species	Locality	Belemnite assemblage zone	N	Mean Riedel-Index	Observed range
G. w. westfalica	West of Bavnodde, beds 1–2	8 (upper Lower Santonian)	46	7.6	3.9–12.0
G. w. westfalica	West of Bavnodde, beds 2–4	8 (upper Lower Santonian)	126	7.4	2.3–12.0
G. w. westfalica	West of Bavnodde, beds 4–7	8 (upper Lower Santonian)	108	7.8	2.8–12.7
G. w. westfalica	East of Bavnodde	7 (upper Lower Santonian)	55	6.5	2.6–9.8
G. w. westfalica	Forchhammers Odde and west of Forchhammers Odde	6 (middle Lower Santonian)	30	6.9	2.7–10.0
G. praewestfalica	Horsemyre Odde and Jydegård	4–5 (lower Lower Santonian)	10	7.8	6.1–9.6
G. praewestfalica	Lägerdorf	Middle–Upper Coniacian	12	7.9	5.9–11.2

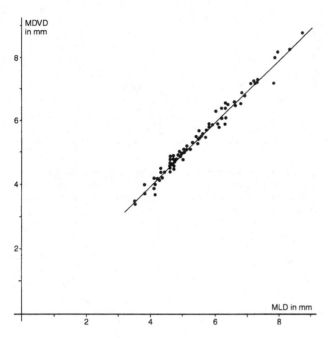

Fig. 34. Scatter plot and regression line of the maximum lateral diameter (MLD) *vs.* the maximum dorsoventral diameter (MDVD) of *Gonioteuthis w. westfalica* from east of Bavnodde. + = mean values.

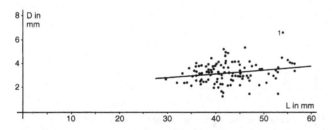

Fig. 36. Scatter plot and regression line of the length of the guard (L) *vs.* the depth of the pseudoalveolus (D) of *Gonioteuthis w. westfalica* from west of Bavnodde, between beds 4 and 7. + = mean values. 1 is MGUH 23723 (Pl. 1:10), which is closely similar to *Goniocamax lundgreni* with respect to the depth of the pseudoalveolus and the cross-section of the anterior end. It is, however, granulated.

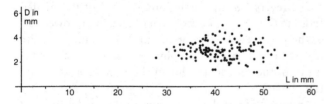

Fig. 38. Scatter plot of the length of the guard (L) *vs.* the depth of the pseudoalveolus (D) of *Gonioteuthis w. westfalica* from west of Bavnodde, between beds 2 and 4. + = mean values. The correlation coefficient is not significant at the 5% level.

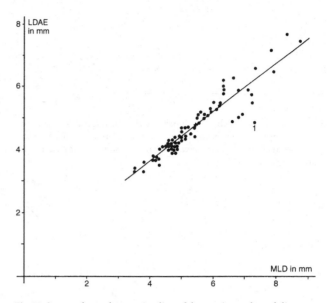

Fig. 35. Scatter plot and regression line of the maximum lateral diameter (MLD) *vs.* the lateral diameter at the alveolar end (LDAE) of *Gonioteuthis w. westfalica* from east of Bavnodde. + = mean values. 1 is MGUH 23720 (Pl. 1:7), which is rather lanceolate in ventral view.

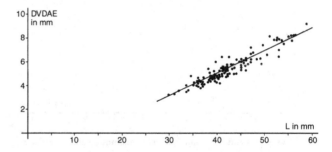

Fig. 37. Scatter plot and regression of the length of the guard (L) *vs.* the dorsoventral diameter at the alveolar end (DVDAE) of *Gonioteuthis w. westfalica* from west of Bavnodde, between beds 4 and 7.

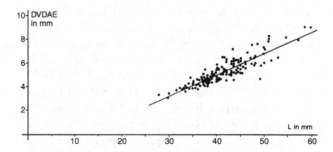

Fig. 39. Scatter plot and regression line of the length of the guard (L) *vs.* the dorsoventral diameter at the alveolar end (DVDAE) of *Gonioteuthis w. westfalica* from west of Bavnodde, between beds 2 and 4.

The means of the Riedel-Index of the samples of *G. praewestfalica* and *G. w. westfalica* from zone 7 are significantly different ($F = 1.9172$, with 54 and 9 degrees of freedom, $P > 0.20$; $t = 2.1764$, with 63 degrees of freedom, $0.05 > P > 0.025$). The means of the Riedel-Index of the samples of *G. w. westfalica* from zone 7 and zone 8 are very highly significantly different ($F = 1.3611$, with 107 and 54 degrees of freedom, $P > 0.10$; $t = 3.9111$, with 161 degrees of freedom, $P < 0.001$).

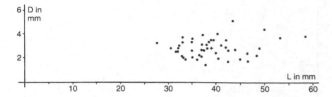

Fig. 40. Scatter plot of the length of the guard (L) *vs.* the depth of the pseudoalveolus (D) of *Gonioteuthis w. westfalica* from west of Bavnodde, between beds 1 and 2. + =mean values. The correlation coefficient is not significant at the 5% level.

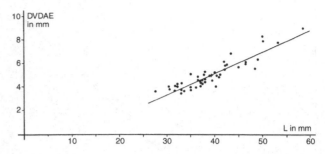

Fig. 41. Scatter plot and regression line of the length of the guard (L) *vs.* the dorsoventral diameter at the alveolar end (DVDAE) of *Gonioteuthis w. westfalica* from west of Bavnodde, between beds 1 and 2. + =mean values.

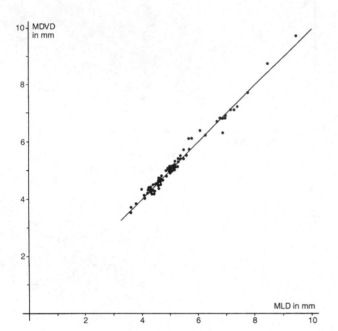

Fig. 42. Scatter plot and regression line of the maximum lateral diameter (MLD) *vs.* the maximum dorsoventral diameter (MDVD) of *Gonioteuthis w. westfalica* from west of Bavnodde, between beds 1 and 2.

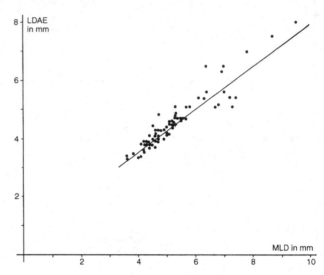

Fig. 43. Scatter plot and regression line of the maximum lateral diameter (MLD) *vs.* the lateral diameter at the alveolar end (LDAE) of *Gonioteuthis w. westfalica* from west of Bavnodde, between beds 1 and 2. + =mean values.

Although the means of the Riedel-Index of the samples of *G. praewestfalica* and *G. w. westfalica* from zone 6 and the means of the samples of *G. w. westfalica* from zone 6 and zone 7 are not statistically significantly different, we consider the trend of the decreasing Riedel-Index as real for the following reasons. (1) It may not be possible to detect statistically significant differences between successive samples of evolutionary lineages, because the differences between closely spaced samples are small (Schulz 1979, p. 33); (2) the means of the Riedel-Index of *G. praewestfalica* from Lägerdorf and Bornholm are very closely similar; (3) the means of the Riedel-Index of *G. w. westfalica* from the lower *westfalica* Zone of Lägerdorf and belemnite assemblage zone 8 of Bornholm are very closely similar; and (4) the mean of the Riedel-Quotient of *G. w. westfalica* from belemnite assemblage zone 7 is significantly smaller than that of *G. praewestfalica* from zones 4–5 and very highly significantly smaller

than that of the oldest sample of *G. w. westfalica* from zone 8.

The reversal of the evolutionary trend of the Riedel-Index in *G. w. westfalica* from zone 7, i.e. in the lower part of the upper Lower Santonian, probably coincides with the start of the well-known phylogenetical increase of the depth of the pseudoalveolus in the *Gonioteuthis* lineage (see above).

An inspection of the histograms of the Riedel-Index of the samples from east of Bavnodde and west of Bavnodde (Fig. 25) shows that most specimens have Riedel-Indices between 4 and 10, and specimens with a very shallow pseudoalveolus (Riedel-Index< 4) occur rarely in all samples. The sample from east of Bavnodde, however, differs from the samples from west of Bavnodde by lacking specimens with a Riedel-Index larger than 10. It is noteworthy that specimens with a Riedel-Index larger than 10 are missing also in the samples from west of Forchhammers

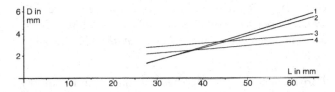

Fig. 44. Regression lines of the length of the guard (L) *vs.* the depth of the pseudoalveolus (D) of samples of *Gonioteuthis praewestfalica* (1) and *G. w. westfalica* (2–4). 1, Jydegård and Horsemyre Odde; 2, west of Forchhammers Odde; 3, east of Bavnodde; and 4, west of Bavnodde, between beds 4 and 7.

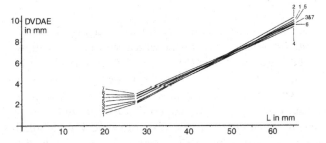

Fig. 45. Regression lines of the length of the guard (L) *vs.* the dorsoventral diameter at the alveolar end (DVDAE) of samples of *Gonioteuthis praewestfalica* (1) and *G. w. westfalica* (2–7). 1, Jydegård and Horsemyre Odde; 2, west of Forchhammers Odde; 3, Forchhammers Odde; 4, east of Bavnodde; 5, west of Bavnodde, between beds 4 and 7; 6, west of Bavnodde, between beds 2 and 4; and 7, west of Bavnodde, between beds 1 and 2.

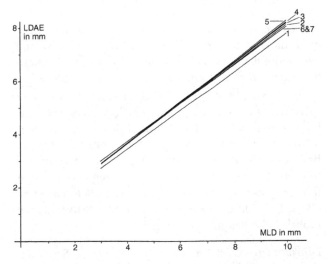

Fig. 46. Regression lines of the maximum lateral diameter (MLD) *vs.* the lateral diameter at the alveolar end (LDAE) of *Gonioteuthis praewestfalica* (1) and *G. w. westfalica* (2–7). 1, Jydegård and Horsemyre Odde; 2, west of Forchhammers Odde; 3, Forchhammers Odde; 4, east of Bavnodde; 5, west of Bavnodde, between beds 4 and 7; 6, west of Bavnodde, between beds 2 and 4; and 7, west of Bavnodde, between beds 1 and 2. Regression lines 2–7 are virtually identical.

Odde and Forchhammers Odde as well as in the sample of *G. praewestfalica*.

The regression lines of the length of the guard *vs.* the depth of the pseudoalveolus of the samples of *G. praewestfalica* and *G. w. westfalica* from west of Forchhammers Odde, east of Bavnodde, and west of Bavnodde, between beds 4 and 7, are shown in Fig. 44. The regression lines of *G. praewestfalica*, and *G. w. westfalica* from west of Forchhammers Odde are closely similar and are situated above the regression lines of *G. w. westfalica* from east of Bavnodde and west of Bavnodde, between beds 4 and 7. The specimens of *G. praewestfalica* and *G. w. westfalica* from west of Forchhammers Odde, therefore, have deeper pseudoalveolus than the specimens of *G. w. westfalica* from east of Bavnodde and west of Bavnodde, between beds 4 and 7 (cf. discussion above of the Riedel-Index).

The regression lines of the length of the guard *vs.* the dorsoventral diameter at the alveolar end of the samples of *G. praewestfalica* and *G. w. westfalica* are shown in Fig. 45, from which is obvious that the samples are virtually identical with respect to these characters.

The means of the Flattening-Quotient of the samples from west of Forchhammers Odde and Forchhammers Odde are slightly larger than the means of the samples from east of Bavnodde and west of Bavnodde. The means of the samples from west of Forchhammers Odde and west of Bavnodde, between beds 4 and 7, are compared with the following result. The *F*-test shows that the variances do not differ significantly at the 5% level. The *t*-test gives a value of 2.4322, with 161 degrees of freedom, which is significant (0.02 > *P* > 0.01). The specimens from west of Forchhammers Odde are thus more flattened ventrally than the specimens from west of Bavnodde, between beds 4 and 7.

With respect to the shape of the guard in ventral view, the following points are worthy of note: (1) most specimens of *G. w. westfalica* are subcylindrical in ventral view, and the MLD/LDAE Index is commonly less than 1.2; (2) because of allometry, adult specimens generally are more lanceolate in ventral view than juvenile specimens (see above); (3) juvenile and adult specimens that are markedly lanceolate in ventral view occur in all samples. These specimens resemble *G. praewestfalica*, but are here assigned to *G. w. westfalica*, because they are not flattened ventrally, some of them are granulated, and they constitute only an insignificant proportion of the samples. They are to be regarded as extreme variants of *G. w. westfalica*. Some of these variants are shown in Pl. 1:7–8.

The regression lines of the maximum lateral diameter *vs.* the lateral diameter at the alveolar end of the samples of *G. praewestfalica* and *G. w. westfalica* are shown in Fig. 46. The regression lines of the six samples of *G. w. westfalica* are virtually identical, whereas the regression line of *G. praewestfalica* is situated below the regression lines of *G. w. westfalica*. The regression line of *G. praewestfalica* and *G. w. westfalica* from west of Bavnodde, between beds 2 and 4, are compared with the following result. The *F*-test of the variances gives a value of 1.3347, with 10 and 155 degrees of freedom, which is not significant ($P > 0.20$). The *t*-test of the slopes gives a value of 0.1177, with 167 degrees of freedom, which is not significant ($P > 0.90$). The *t*-test of the positions gives a value of 2.0772, with 167 degrees of freedom, which is significant ($0.05 > P > 0.025$). *G. praewestfalica* is thus significantly more lanceolate in ventral view than *G. w. westfalica*.

In conclusion, we regard the specimens of *G. w. westfalica* from west of Forchhammers Odde and Forchhammers Odde as early forms of *G. w. westfalica*, which are transitional to *G. praewestfalica*, because they are somewhat flattened ventrally. Moreover, specimens with Riedel-Indices larger than 10 do not occur at these sites. The specimens from east of Bavnodde are considered to be slightly younger than those from west of Forchhammers Odde and Forchhammers Odde, because generally they are not flattened ventrally. On the other hand, the specimens from east of Bavnodde are considered to be slightly older than those from west of Bavnodde, because generally they have a more shallow pseudoalveolus. This is consistent with the stratigraphical results obtained by the structural geology and the range of other belemnites (Fig. 12).

The affinity to *G. ernsti* sp.nov. is discussed below. *G. westfalica mujnakensis* Naidin, 1964, is stouter and more lanceolate in ventral view than *G. w. westfalica*. In addition, *G. westfalica mujnakensis* is markedly flattened ventrally. *G. westfalica aralensis* Naidin, 1964, is stouter and more flattened ventrally than *G. w. westfalica*. Moreover, it differs by the structure of the alveolar end. *G. w. westfalica* is distinguished from specimens of *Goniocamax lundgreni* with a shallow pseudoalveolus by the shape of the bottom of the ventral fissure. In addition, *G. w. westfalica* is smaller than *G. lundgreni*.

Distribution. – *G. w. westfalica* occurs commonly in NW Germany (Ernst 1964, 1968; Ernst & Schulz 1974), southern Scandinavia (Christensen 1973, 1975a, 1986, 1993a), and at Lonzée in SE Belgium (Christensen 1994). Outside this area it has been recorded from most parts of the Central European Subprovince except east of Ukraine. It occurs in the upper Lower and lower Middle Santonian.

Gonioteuthis ernsti sp.nov.

Pl. 1:12–13

Synonymy. – □1964 *Gonioteuthis* aff. *westfalica non* Birkelund – Ernst, Pl. 2:1; Pl. 3:11. □?1964 *Gonioteuthis lundgreni/westfalica* – Ernst, Pl. 3:7.

Types. – The holotype and five paratypes are from the Bavnodde Greensand, west of Bavnodde. Holotype is

GPIK 3905, between beds 4 and 5. Paratypes are GM 1995.1-2, between beds 5 and 6; GIPK 3922, loose; GPIK 3923, between beds 4 and 5; and MGUH 23725, between beds 3 and 5.

Derivation of name. – The species is named in honour of Professor Gundolf Ernst, who first recognized this form.

Dimensions. – See Table 31.

Diagnosis. – Guard very slender and usually slightly flattened ventrally; lanceolate in ventral view and lanceolate or cylindrical in lateral view; anterior end conical with a central, very shallow pseudoalveolus; guard with dorsolateral double furrows, single lateral furrows, and longitudinal striae.

Description. – Guard small and usually slightly flattened ventrally; anterior end conical with a central, very shallow pseudoalveolus (Riedel-Index ca. 4–6, mean value 4.7; Riedel-Quotient ca. 17–27, mean value ca. 22); anterior end with dorsal and ventral embayments; guard lanceolate in ventral view, with largest diameter in middle part or posterior third, and lanceolate or cylindrical in lateral view; apex with indistinctly separated mucro, slightly displaced towards dorsal side; guard very slender (Slenderness-Quotient ca. 9–1); cross-section of anterior end subtriangular and compressed laterally; short ventral fissure or ventral furrow present in some specimens; dorsolateral longitudinal depressions and double furrows, single lateral furrows, and longitudinal striae present; otherwise smooth guard.

Biometry. – The types were analyzed by univariate statistical methods (Table 32).

Discussion. – *G. ernsti* is closely similar to *G. w. westfalica*, but is more slender, more lanceolate in ventral view, and has a more shallow pseudoalveolus. *G. ernsti* is more slender and less flattened ventrally than *G. praewestfalica*.

Ernst (1964, Pl. 2:1; Pl. 3:11) figured two very slender specimens with a very shallow pseudoalveolus from the lower *westfalica* Zone of the Münster Basin as *G.* aff. *westfalica non* Birkelund. The specimens from the Münster Basin differ in no significant respect from the specimens from Bornholm, and they are here referred to *G. ernsti*. A third specimen, figured as *G. lundgreni/westfalica* by Ernst (1964, Pl. 3:7), has a slightly deeper pseudoalveolus but probably also belongs to *G. ernsti*

Distribution. – *G. ernsti* is a very rare species. Six specimens are recorded here from the lower third of the Bavnodde Grensand exposed west of Bavnodde, which is upper Lower Santonian. Two, possibly three, specimens are known from the Münster Basin, lower *westfalica* beds (=*undulatoplicatus/cordiformis* boundary beds; Ernst 1964), i.e. upper Lower Santonian.

Table 31. Dimensions of *Gonioteuthis ernsti* sp.nov. from the Bavnodde Greensand. GPIK 3905, holotype; MGUH 23725; paratype. *=estimated. Measurements in millimetres.

Character	GPIK 3905	MGUH 23725
L	53.0	51*
D	2.5	1.9
DVDAE	6.0	4.8
LDAE	5.8	4.5
MLD	6.6	6.7
MDVD	6.3	6.8
RQ	21.2	25.5
RI	4.7	3.9
SQ	8.8	10.9
L/MLD	8.0	7.6
FQ	1.05	0.99
MLD/LDAE	1.14	1.49
DVDAE/LDAE	1.03	1.07

Table 32. Univariate analysis of *Gonioteuthis ernsti* sp.nov. from the Bavnodde Greensand, west of Bavnodde. Measurements in millimetres.

Character	N	X̄	SD	CV	OR
L	5	49.2	3.0	6.2	45.0–53.0
D	6	2.3	0.4	16.7	1.8–2.8
DVDAE	6	5.5	1.0	18.7	4.8–7.3
LDAE	6	5.2	1.0	19.4	4.5–7.0
MLD	6	6.6	1.1	17.4	5.0–8.5
MDVD	6	6.4	1.0	16.2	5.0–8.1
RQ	5	22.0	4.0	18.1	17.1–26.8
RI	5	4.7	0.9	18.1	3.7–5.8
SQ	5	9.7	0.8	8.2	8.8–10.6
L/MLD	5	8.0	0.6	7.7	7.4–9.0
FQ	6	1.02	0.03	2.5	0.99–1.05
MLD/LDAE	6	1.27	0.15	11.8	1.09–1.49
DVDAE/LDAE	6	1.05	0.02	1.6	1.03–1.07

Genus *Goniocamax* Naidin, 1964

Type species. – *Actinocamax lundgreni* Stolley, 1897 by orginal designation.

Emended diagnosis. – Small to medium-sized, non granulated belemnitellids with a shallow to deep pseudoalveolus (Riedel-Index generally 8–20); guard markedly flattened ventrally, lanceolate in ventral view and subcylindrical in lateral view; juvenile guard short and stout; guard with dorsolateral longitudinal depressions and double furrows, vascular imprints, and longitudinal striae; longitudinal striae usually more distinct than vascular markings; Schatzky distance small, 2–4 mm; bottom of ventral fissure commonly straight or slightly curved, forming a medium-sized angle, ca. 30–50°, with the wall of the pseudoalveolus; vascular imprints branch off the

dorsolateral double furrows posteriorly at an angle less than 30°.

Discussion. – *Goniocamax* was established by Naidin (1964b) as a subgenus of *Gonioteuthis* (see above). It is here raised to the rank of a genus, following the suggestion by Ernst & Schulz (1974). We include the type species and its close relatives in the genus, i.e. *G. birkelundae* sp.nov., *G. striatus* sp.nov., *Actinocamax esseniensis* Christensen, 1982, and *Belemnitella mirabilis* Arkhangelsky, 1912.

B. mirabilis from northern Kazakhstan was established by Arkhangelsky (1912, Pl. 10:28–30) and refigured by Jeletzky (1949, Fig. 4). It was later considered a subspecies of *B. propinqua* by Naidin (1964a, b). The holotype, by monotypy, is the original of Arkhangelsky, who reported the following measurements: L: 68 mm; D: 13 mm; DVDAE: 8 mm; LDAE: 8 mm; MLD: 10.8 mm; and MDVD: 9.7 mm. On the basis of these measurements the following indices are calculated: RI: 19.1; RQ: 5.2; SQ: 8.5; L/MLD: 6.3; FQ: 1.1; and MLD/LDAE: 1.4. The Schatzky distance is ca. 3 mm, the fissure angle is ca. 33°, and the pseudoalveolus is covered by conellae (Jeletzky 1949, Fig. 4). Arkhangelsky also noted that the surface markings mainly consist of striae.

This taxon should be placed in the genus *Goniocamax* owing to its size and shape of the guard, depth of pseudoalveolus, slenderness, and surface markings. The affinity to species described here is discussed below. Arkhangelsky (1912) noted that it was found together with *A. verus fragilis* and *Belemnitella praecursor*. These taxa co-occur in the Upper Santonian and Lower Campanian elsewhere. The age of *G. mirabilis* is most likely incorrect, since *Goniocamax* is not recorded from the Upper Santonian or Lower Campanian.

Goniocamax is closely similar to *Gonioteuthis*, but differs from that genus by being non-granulated, flattened ventrally, and lanceolate in ventral view. It is worthy of note, however, that Naidin (1964b, p. 135) recorded a single granulated specimen of *G. lundgreni excavata* (Sinzow) (no. 8012/4).

The earliest member of the *Gonioteuthis* lineage, *G. praewestfalica*, usually is not granulated, and the guard is flattened ventrally and club-shaped in ventral view. In the succeeding taxon, *G. w. westfalica*, about half the specimens are granulated, and extreme variants are markedly flattened ventrally and club-shaped in ventral view. Variants of *Goniocamax lundgreni* with a very shallow pseudoalveolus are therefore difficult to distinguish from variants of *Gonioteuthis praewestfalica* with a pseudoalveolus and non-granulated extreme variants of *G. w. westfalica*. However, the bottom of the ventral fissure is usually sine-shaped and forms a large angle, often more than 90°, with the wall of the pseudoalveolus in *G. praewestfalica* and *G. westfalica* (Ernst 1964, Figs. 16, 22), whereas it is generally straight or slightly curved and forms a medium-sized angle, ca. 30–50°, with the wall of the pseudoalveolus in *G. lundgreni* (Fig. 16A). In addition, *G. lundgreni* is larger than *G. praewestfalica* and *G. w. westfalica*.

Naidin (1964b) suggested an evolutionary lineage, including *G. lundgreni excavata* (Sinzow, 1915), *G. lundgreni postexcavata* Naidin, 1964b, *Belemnitella propinqua propinqua* (Moberg, 1885), and *B. propinqua rylskiana* Nikitin, 1958. These taxa are discussed below. The evolutionary trends of the *Goniocamax* group and *Belemnitella* lineage are discussed above.

Distribution. – *Goniocamax* is mainly distributed in the Central Russian Subprovince and Baltoscandia. A few representatives are recorded from the Central European Subprovince. It occurs in the Coniacian and Lower Santonian.

Goniocamax lundgreni lundgreni (Stolley, 1897)

Pl. 2:1–5, 7–12; Pl. 3:1; Fig. 16A

Synonymy. – □1897 *Actinocamax Lundgreni* – Stolley, p. 285, Pl. 3:16–20 (*non* Pl. 3:15 = *Actinocamax verus* cf. *antefragilis*). □1897 *Actinocamax mammillatus* mut. (*ant.*) *bornholmensis* – Stolley, p. 288, Pl. 4:1. □1897 *Actinocamax propinquus* Moberg *mut.* (*var.*) *nov.* – Stolley, p. 295, Pl. 3:23. □1912 *Actinocamax propinquus* Moberg – Arkhangelsky, p. 585, Pl. 10:14–15, ?*non* 23–27, 34–36. □1912 *Actinocamax intermedius* sp. n. – Arkhangelsky, p. 582, Pl. 9:30–31; Pl. 10:6, 16–18, ?*non* 27. □1915 *Actinocamax plenus* Miller var. *excavata* – Sinzow, p. 144, Pl. 8:14–17, ?*non* fig. 27. □1918 *Actinocamax bornholmensis* Stolley – Ravn, p. 33, Pl. 2:7. □1918 *Actinocamax* sp. (cf. *Act. strehlenensis* Fritsch) – Ravn, p. 34, Pl. 2:8. □1946 *Actinocamax lundgreni* Stolley – Ravn, p. 30. □1957 *Actinocamax lundgreni lundgreni* Stolley – Birkelund, p. 13, Pl. 1:5–6. □1957 *Actinocamax lundgreni excavata* Sinzow – Birkelund, p. 18, Pl. 1:7–8. □1957 *Actinocamax lundgreni lundgreni*? Stolley – Birkelund, p. 14. □1957 *Actinocamax lundgreni excavata*? Sinzow – Birkelund, p. 19. □ pars 1957 *Actinocamax westfalicus* (Schlüter) – Birkelund, p. 27. □1957 *Actinocamax* aff. *westfalicus* (Schlüter) – Birkelund, pp. 27–28, Pl. 2:3. □1958 *Actinocamax intermedius* Arkhangelsky – Nikitin, p. 5, Pl. 1:4–8. □1958 *Actinocamax propinquus* Moberg – Nikitin, p. 12, Pl. 1:9–15; Pl. 3:7. □1964 *Gonioteuthis lundgreni*/aff. *westfalica sensu* Birkelund – Ernst, p. 161, Pl. 3:5–6. □ 1964b *Gonioteuthis* (*Goniocamax*) *lundgreni lundgreni* (Stolley) – Naidin, p. 127, Pl. 7:5–7. □1964b *Gonioteuthis* (*Goniocamax*) *lundgreni excavata* (Sinzow) – Naidin, p. 133, Pl. 7:8. □1964b *Gonioteuthis* (*Goniocamax*) *lundgreni postexcavata* – Naidin, p. 135. □1972 *Actinocamax* (*Actinocamax*) *propinquus propinquus* Moberg – Glazunova, p. 106, Pl. 45:1–5.

□1972 *Actinocamax* (*Actinocamax*) aff. *propinquus propinquus* Moberg – Glazunova, p. 107, Pl. 46:1. □1972 *Gonioteuthis* (*Goniocamax*) cf. *lündgreni lündgreni* (Stolley) – Glazunova, p. 113, Pl. 46:3–4. □1973 *Gonioteuthis lundgreni* (Stolley) – Christensen, p. 31, Pl. 10:6–9. □1974 *Gonioteuthis* (*Goniocamax*) *lundgreni lundgreni* (Stolley) – Naidin, p. 211, Pl. 73:8. □1974 *Gonioteuthis* (*Goniocamax*) *lundgreni excavata* (Sinzow) – Naidin, p. 211, Pl. 73:9. □1975a *Actinocamax lundgreni* Stolley – Christensen, p. 28. □1982 *Actinocamax lundgreni* Stolley – Christensen, p. 76. □1986 '*Actinocamax*' *lundgreni* Stolley – Christensen, p. 30. □1991 '*Actinocamax*' *lundgreni* Stolley – Christensen, p. 734, Pl. 4:1–6.

Type. – Lectotype, by subsequent designation of Birkelund (1957, p. 4), MGUH 1517, the original of Stolley (1897, Pl. 3:18), from the marl at Muleby Å, Bornholm. This marl is upper Lower Coniacian (see above).

Material. – Arnager Limestone Formation: 37 specimens from the Arnager Limestone west of Arnager, including MGUH 1690-1, figured by Ravn (1918, Pl. 2:7–8); MGUH 7837-8, figured by Birkelund (1957, Pl. 1:5–6); and the holotype of *Actinocamax mammillatus* mut. *bornholmensis* Stolley, GPIG unregistered, figured by Stolley (1897, Pl. 4:1). Precisely located specimens came from the bottom conglomerate of the limestone, 50 cm, 120 cm, and 150–250 cm above the base, and the very top of the formation exposed in the bay between Arnager and Horsemyre Odde. 20 specimens from the marl at Muleby Å, loc. 1, 3, 5, and 8, including MGUH 1515-9, figured by Stolley (1897, Pl. 3:16–20). 24 specimens from the marl at Stampe Å, most of which came from loc. 5 (Fig. 4).

Bavnodde Greensand Formation: seven specimens from east of Horsemyre Odde; 76 specimens from east of Forchhammers Odde; 17 specimens from between Forchhammers Odde and Skidteper; 21 specimens from west of Skidteper; six specimens from Blykobbe Å at Risenholm; 74 specimens from Jydegård, including MGUH 7839-40, figured as *A. lundgreni excavata* by Birkelund (1957, Pl. 1:7–8), and MGUH 7843, figured as *A.* aff. *westfalicus* by Birkelund (1957, Pl. 2:3); three specimens from west of Horsemyre Odde; 45 specimens from Horsemyre Odde; two specimens from west of Forchhammers Odde; three specimens from Forchhammers Odde.

Dimensions. – See Table 33.

Description. – Guard medium-sized and flattened ventrally (Flattening-Quotient 0.97–1.19, mean values 1.04–1.10), lanceolate in ventral view, with largest diameter in the middle part or posterior third, and slightly lanceolate or subcylindrical in lateral view; anterior end with a shallow pseudoalveolus (Riedel-Index 8–17, mean values 11–15; Riedel-Quotient usually 5–14, mean values ca. 7–9); apex with indistinctly separated mucro; Slenderness-Quotient generally 6–8; adult specimens usually stouter

Coniacian and Santonian belemnites 43

Table 33. Dimensions of *Goniocamax l. lundgreni* from the Arnager Limestone Formation. MGUH 1517, lectotype; 1, holotype of *Actinocamax mammillatus* var. *bornholmensis* Stolley, GPIG unregistered. *=estimated. Measurements in millimetres.

Character	MGUH 1517	1
L	62*	57*
D	5.5	7.2
DVDAE	8.1	7.7
LDAE	8.3	7.2
MLD	10.9	8.4
MDVD	9.6	8.2
LVF	–	–
RQ	11.3	7.9
RI	8.9	12.6
SQ	7.7	7.4
L/MLD	5.7	6.8
FQ	1.14	1.02
MLD/LDAE	1.31	1.17
DVDAE/LDAE	0.98	1.02

than juvenile specimens because of allometric growth; cross-section of anterior end generally subtriangular and compressed laterally; wall of pseudoalveolus almost straight or curved, often with conellae; pseudoalveolus often with dorsal and ventral notches; short ventral fissure may be present; dorsolateral longitudinal depressions and double furrows well developed; surface of well preserved guards with vascular imprints and distinct longitudinal striae; striae more distinct than vascular imprints; Schatzky distance small, 2–4 mm; bottom of ventral fissure usually straight or slightly curved forming a medium-sized angle, ca. 30–50°, with the wall of the pseudoalveolus.

Biometry. – Eight samples are analyzed by univariate and bivariate methods. Three samples are from the Arnager

Table 34. Univariate analysis of *Goniocamax l. lundgreni* from the Arnager Limestone Formation, west of Arnager. Measurements in millimetres.

Character	N	X̄	SD	CV	OR
L	11	55.0	6.7	12.2	37.8–63.8
D	19	5.6	1.4	24.8	3.2–8.2
DVDAE	18	7.5	1.6	21.5	4.8–10.8
LDAE	18	7.0	1.5	21.3	4.5–10.8
MLD	31	8.6	1.7	19.1	5.8–13.8
MDVD	34	7.9	1.4	18.1	5.4–12.6
LVF	6	2.2	0.8	39.0	0.8–3.0
RQ	11	9.2	1.2	12.5	7.4–11.3
RI	11	11.0	1.4	12.4	8.8–13.4
SQ	10	7.8	0.9	11.5	6.4–9.6
L/MLD	11	6.4	0.5	7.3	5.8–7.4
FQ	31	1.09	0.04	4.0	1.01–1.19
MLD/LDAE	16	1.24	0.09	7.3	1.06–1.39
DVDAE/LDAE	17	1.04	0.05	4.9	0.94–1.11

Table 35. Univariate analysis of *Goniocamax l. lundgreni* from the marl of the Arnager Limestone Formation, Muleby Å. Measurements in millimetres. *MGUH 1516 is not included, because it is exfoliated anteriorly. This specimen is markedly lanceolate in ventral view (MLD/LDAE=1.61) because of the exfoliation.

Character	N	X̄	SD	CV	OR
L	7	56.0	7.6	13.5	48.1–65.5
D	11	6.4	2.3	36.1	1.7–9.8
DVDAE	10	7.7	1.4	17.9	6.2–10.1
LDAE	11	7.2	1.4	19.3	5.4–9.9
MLD	11	8.7	1.7	19.0	6.8–11.9
MDVD	11	7.5	2.3	34.4	6.3–10.7
RQ*	6	8.1	1.7	21.4	6.7–11.3
RI*	6	12.7	2.3	18.4	8.9–15.0
SQ	6	7.3	0.6	8.3	6.5–7.8
L/MLD	7	6.0	0.4	7.3	5.5–6.5
FQ	11	1.09	0.04	3.7	1.00–1.14
MLD/LDAE	8	1.29	0.14	10.7	1.19–1.61
DVDAE/LDAE	10	1.05	0.05	4.5	0.98–1.11

Table 36. Univariate analysis of *Goniocamax l. lundgreni* from the marl of the Arnager Limestone Formation, Stampe Å. Measurements in millimetres.

Character	N	X̄	SD	CV	OR
L	9	53.1	8.4	15.8	43.0–67.0
D	17	7.4	2.0	26.8	4.0–10.0
DVDAE	14	7.4	1.7	22.4	4.6–11.0
LDAE	16	7.0	1.5	21.7	4.5–10.6
MLD	17	9.1	2.1	23.2	6.2–13.5
MDVD	18	8.3	1.9	22.6	5.7–12.3
RQ	9	7.1	1.7	24.2	5.0–10.7
RI	9	14.8	3.2	21.6	9.3–20.0
SQ	8	6.9	0.8	10.9	6.1–8.3
L/MLD	9	6.1	0.8	13.2	5.0–7.3
FQ	18	1.10	0.03	2.8	1.05–1.15
MLD/LDAE	12	1.24	0.09	7.5	1.09–1.40
DVDAE/LDAE	14	1.06	0.07	6.4	0.93–1.17

Table 37. Univariate analysis of *Goniocamax l. lundgreni* from the Bavnodde Greensand, east of Forchhammers Odde. Measurements in millimetres.

Character	N	X̄	SD	CV	OR
L	42	43.8	5.3	12.2	34.5–55.3
D	62	5.1	1.3	25.7	2.5–8.1
DVDAE	63	6.1	1.4	23.0	3.7–10.4
LDAE	59	5.8	1.3	21.9	3.8–9.6
MLD	69	6.9	1.3	19.3	4.5–11.0
MDVD	69	6.6	1.3	19.7	4.4–10.8
RQ	39	8.6	1.8	21.1	6.0–13.9
RI	39	12.1	2.4	19.9	7.2–16.7
SQ	36	not calculated			5.3–8.3
L/MLD	40	not calculated			5.3–7.7
FQ	67	1.05	0.02	2.3	0.99–1.10
MLD/LDAE	53	1.18	0.08	7.2	1.04–1.45
DVDAE/LDAE	60	1.05	0.04	4.0	0.93–1.12

Table 38. Univariate analysis of *Goniocamax l. lundgreni* from the Bavn-odde Greensand, between Forchhammers Odde and Skidteper. Measurements in millimetres.

Character	N	X̄	SD	CV	OR
L	12	45.1	3.4	7.5	39.9–51.5
D	13	5.4	1.0	18.9	3.8–7.1
DVDAE	14	6.5	1.0	15.7	4.9–8.5
LDAE	14	6.1	0.8	13.3	4.9–7.8
MLD	17	7.1	1.2	16.7	5.3–9.2
MDVD	17	6.8	1.1	16.2	5.1–8.9
RQ	12	8.4	1.0	12.2	6.5–9.7
RI	12	12.2	1.6	12.8	10.3–13.8
SQ	12	6.9	0.6	8.9	5.6–7.5
L/MLD	12	6.2	0.7	11.0	5.0–7.0
FQ	16	1.04	0.02	2.1	1.00–1.09
MLD/LDAE	14	1.21	0.07	5.6	1.11–1.32
DVDAE/LDAE	14	1.07	0.03	2.9	1.00–1.12

Table 39. Univariate analysis of *Goniocamax l. lundgreni* from the Bavn-odde Greensand, west of Skidteper. Measurements in millimetres.

Character	N	X̄	SD	CV	OR
L	16	44.0	6.1	13.8	34.5–56.2
D	18	5.7	1.3	22.7	4.0–7.7
DVDAE	17	6.0	1.3	22.0	4.5–8.7
LDAE	17	5.7	1.2	20.4	4.4–8.1
MLD	20	6.7	1.4	20.4	5.0–9.3
MDVD	20	6.4	1.3	20.8	4.7–9.1
RQ	14	7.6	1.1	14.5	5.8–9.4
RI	14	13.3	1.9	14.3	10.8–17.2
SQ	13	not calculated			6.1–8.7
L/MLD	16	not calculated			5.6–7.6
FQ	19	1.05	0.03	3.2	1.00–1.17
MLD/LDAE	15	1.14	0.07	5.7	1.06–1.27
DVDAE/LDAE	16	1.05	0.03	2.9	0.98–1.08

Table 40. Univariate analysis of *Goniocamax l. lundgreni* from the Bavn-odde Greensand, Jydegård. Measurements in millimetres.

Character	N	X̄	SD	CV	OR
L	34	47.2	9.4	20.0	28.0–67.9
D	64	5.2	1.4	26.6	2.0–8.7
DVDAE	64	6.1	1.6	26.9	2.7–9.8
LDAE	61	5.9	1.6	26.4	2.8–9.4
MLD	52	7.3	2.1	28.2	3.2–11.4
MDVD	56	7.0	1.9	27.2	3.3–10.6
RQ	34	8.8	1.6	18.1	6.4–13.0
RI	34	11.6	1.9	16.4	7.7–15.6
SQ	36	not calculated			6.0–9.7
L/MLD	36	not calculated			5.2–8.4
FQ	52	1.05	0.03	3.0	0.97–1.13
MLD/LDAE	40	1.23	0.11	8.7	1.00–1.47
DVDAE/LDAE	60	1.05	0.04	4.2	0.96–1.16

Table 41. Univariate analysis of *Goniocamax l. lundgreni* from the Bavn-odde Greensand, Horsemyre Odde. Measurements in millimetres.

Character	N	X̄	SD	CV	OR
L	19	47.8	7.3	15.3	37.0–60.0
D	38	5.2	1.6	29.8	3.0–9.5
DVDAE	38	6.5	1.5	22.8	4.4–9.7
LDAE	38	6.1	1.3	21.6	4.1–9.0
MLD	35	7.2	1.7	23.2	4.9–10.9
MDVD	35	6.9	1.6	23.2	4.7–10.7
RQ	17	8.7	1.6	18.0	6.0–11.9
RI	17	11.9	2.1	17.8	8.4–16.7
SQ	17	not calculated			5.7–7.6
L/MLD	19	not calculated			5.1–6.9
FQ	35	1.04	0.02	2.3	1.00–1.08
MLD/LDAE	28	1.18	0.07	6.2	1.08–1.40
DVDAE/LDAE	37	1.06	0.05	4.8	0.90–1.14

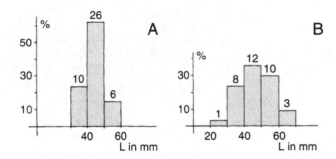

Fig. 47. Histograms of the length of the guard (L) in mm of two samples of *Goniocamax l. lundgreni.* □A. East of Forchhammers Odde. □B. Jydegård. The figures above the bars are the actual number of specimens.

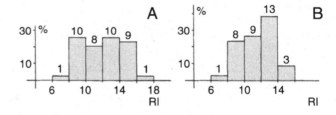

Fig. 48. Histograms of the Riedel-Index (RI) of two samples of *Goniocamax l. lundgreni.* □A. East of Forchhammers Odde. □B. Jydegård. The figures above the bars are the actual number of specimens.

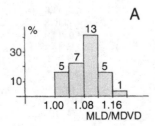

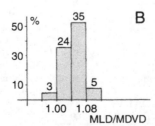

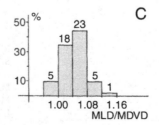

Fig. 49. Histograms of the Flattening-Quotient (MLD/MDVD) of three samples of *Goniocamax l. lundgreni.* □A. Arnager. □B. East of Forchhammers Odde. □C. Jydegård. The figures above the bars are the actual number of specimens.

Limestone Formation: (1) 34 specimens from the Arnager Limestone west of Arnager, (2) 11 specimens from the marl at Muleby Å, and (3) 18 specimens from the marl at Stampe Å. Five samples are from the Bavnodde Greensand Formation: (1) 69 specimens from east of Forchhammers Odde, (2) 17 specimens from between Forchhammers Odde and Skidteper, (3) 20 specimens from west of Skidteper, (4) 64 specimens from Jydegård, and (5) 38 specimens from Horsemyre Odde.

Univariate analysis. – The results of the univariate analyses are shown in Tables 34–41. Histograms of the length of the guard, Riedel-Index, Flattening-Quotient, and MLD/LDAE Index of the samples from east of Forchhammers Odde and Jydegård are shown in Figs. 47–50. Fig. 49 also includes the histogram of the Flattening-Quotient of the sample from Arnager.

Bivariate analyses. – The following regression analyses are made: (1) the length of the guard *vs.* the depth of the pseudoalveolus, (2) the length of the guard *vs.* the dorsoventral diameter at the alveolar end, (3) the length of the guard *vs.*

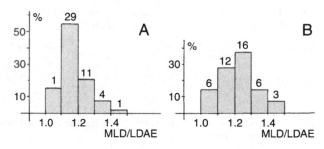

Fig. 50. Histograms of the MLD/LDAE Index of two samples of *Goniocamax l. lundgreni*. □A. East of Forchhammers Odde. □B. Jydegård. The figures above the bars are the actual number of specimens.

the maximum lateral diameter, (4) the maximum lateral diameter *vs.* the maximum dorsoventral diameter, and (5) the maximum lateral diameter *vs.* the lateral diameter at the alveolar end. The equations of the regressions lines are given in Tables 42–49, and the scatter plots are shown in Figs. 51–66.

Table 42. Estimates of statistical parameters of five regression analyses of *Goniocamax l. lundgreni* from the Arnager Limestone, west of Arnager. Measurements in millimetres.

$y = a + bx$	N	r	Probability	SD_a	SD_b	SD_{yx}	t_a	Probability
D = −2.5410 + 0.1572L	11	0.8297	0.01 > P > 0.001	1.9506	0.0352	0.7486	1.3027	0.25 > P > 0.20
DVDAE = −2.6125 + 0.1783L	10	0.8651	0.01 > P > 0.001	2.0287	0.0363	0.7573	1.2878	0.25 > P > 0.20
MLD = −1.8814 + 0.1925L	11	0.9098	P < 0.001	1.5737	0.0284	0.6040	1.1956	0.30 > P > 0.25
MDVD = 0.3098 + 0.8750MLD	31	0.9699	P < 0.001	0.3526	0.0401	0.3620	0.8786	0.40 > P > 0.30
LDAE = −0.3951 + 0.8471MLD	15	0.9274	P < 0.001	0.8697	0.0999	0.5156	0.4543	0.70 > P > 0.60

Table 43. Estimates of statistical parameters of two regression analyses of *Goniocamax l. lundgreni* from the marl of the Arnager Limestone Formation, Muleby Å. Measurements in millimetres.

$y = a + bx$	N	r	Probability	SD_a	SD_b	SD_{yx}	t_a	Probability
D = −4.2065 + 0.1948L	7	0.6815	0.10 > P > 0.05	(no significant correlation)				
DVDAE = −3.1909 + 0.1947L	6	0.9316	0.01 > P > 0.001	5.3387	0.0924	0.6684	0.5977	0.50 > P > 0.40

Table 44. Estimates of statistical parameters of two regression analyses of *Goniocamax l. lundgreni* from the sandy marl of the Arnager Limestone Formation, Stampe Å. Mesasurements in mm.

$y = a + bx$	N	r	Probability	SD_a	SD_b	SD_{yx}	t_a	Probability
D = −1.6481 + 0.1795L	9	0.7074	0.05 > P > 0.02	3.6411	0.0679	1.6091	0.4526	0.70 > P > 0.60
DVDAE = −3.1935 + 0.2065L	8	0.9504	P < 0.001	1.5800	0.0291	0.6673	2.0212	0.10 > P > 0.05

Table 45. Estimates of statistical parameters of five regression analyses of *Goniocamax l. lundgreni* from the Bavnodde Greensand, east of Forchhammers Odde. Measurements in millimetres.

$y = a + bx$	N	r	Probability	SD_a	SD_b	SD_{yx}	t_a	Probability
D = −2.2259 + 0.1725L	39	0.6766	P < 0.001	1.5493	0.0349	1.1636	1.4367	0.20 > P > 0.10
DVDAE = −3.3777 + 0.2182L	36	0.8925	P < 0.001	0.8367	0.0189	0.5907	4.0370	P < 0.001
MLD = −2.4221 + 0.2136L	41	0.9038	P < 0.001	0.7439	0.0168	0.5721	3.2562	0.005 > P > 0.001
MDVD = −0.0501 + 0.9610MLD	69	0.9899	P < 0.001	0.1157	0.0164	0.1810	0.4330	0.70 > P > 0.60
LDAE = 0.0499 + 0.8417MLD	53	0.9463	P < 0.001	0.2713	0.0383	0.3921	0.1839	0.90 > P > 0.80

Table 46. Estimates of statistical parameters of two regression analyses of *Goniocamax l. lundgreni* from the Bavnodde Greensand, between Forchhammers Odde and Skidteper. Measurements in millimetres.

$y = a + bx$	N	r	Probability	SD_a	SD_b	SD_{yx}	t_a	Probability
D=−3.4126+0.1974L	12	0.7149	0.01>P>0.001	2.7900	0.0618	0.6972	1.2232	0.25>P>0.20
DVDAE=−4.1221+0.2373L	12	0.8112	0.01>P>0.001	2.4918	0.0551	0.6215	1.6543	0.20>P>0.10

Table 47. Estimates of statistical parameters of two regression analyses of *Goniocamax l. lundgreni* from the Bavnodde Greensand, west of Skidteper. Measurements in millimetres.

$y = a + bx$	N	r	Probability	SD_a	SD_b	SD_{yx}	t_a	Probability
D=−1.9707+0.1790L	14	0.8336	P<0.001	1.5546	0.0350	0.7996	1.2677	0.25>P>0.20
DVDAE=−3.0375+0.2100L	13	0.9618	P<0.001	0.7465	0.0167	0.3795	4.0689	0.005>P>0.001

Table 48. Estimates of statistical parameters of five regression analyses of *Goniocamax l. lundgreni* from the Bavnodde Greensand, Jydegård. Measurements in millimetres.

$y = a + bx$	N	r	Probability	SD_a	SD_b	SD_{yx}	t_a	Probability
D=0.1067+0.1140L	34	0.7711	P<0.001	0.8078	0.0168	0.9105	0.1321	0.90>P>0.80
DVDAE=−1.9873+0.1785L	34	0.9351	P<0.001	0.5813	0.0121	0.6551	3.4190	0.005>P<0.001
MLD=−2.6852+0.2166L	34	0.9574	P<0.001	0.5582	0.0116	0.6282	4.8105	P<0.001
MDVD=0.2579+0.9158MLD	52	0.9959	P<0.001	0.0708	0.0093	0.1370	3.6427	P<0.001
LDAE=0.5014+0.7414MLD	43	0.9603	P<0.001	0.2564	0.0332	0.4567	1.9559	0.10>P>0.05

Table 49. Estimates of statistical paramaters of four regression analyses of *Goniocamax l. lundgreni* from the Bavnodde Greensand, Horsemyre Odde. Measurements in millimetres.

$y = a + bx$	N	r	Probability	SD_a	SD_b	SD_{yx}	t_a	Probability
D=−2.6798+0.1748L	17	0.7837	P<0.001	1.6554	0.0337	0.9626	1.6189	0.20>P>0.10
DVDAE=−1.9271+0.1889L	17	0.9473	P<0.001	0.9008	0.0183	0.5238	2.1392	0.05>P>0.025
MDVD=0.0094+0.9596MLD	35	0.9949	P<0.001	0.2281	0.0310	0.3004	0.0412	P>0.90
LDAE=0.5105+0.7820MLD	28	0.9669	P<0.001	0.2701	0.0353	0.3075	1.8899	0.10>P>0.05

The correlation coefficients are tested for significance, and the *y*-intercepts are tested by *t*-tests to see if they differ significantly from zero (Tables 42–49).

The coefficients of determination vary from 45 to 70% in the analyses of the length of the guard *vs.* the depth of the pseudoalveolus. They are larger in the analyses of the length of the guard *vs.* the dorsoventral diameter at the alveolar end (65–90%), and the maximum lateral diameter *vs.* the lateral diameter at the alveolar end (85–95%). They are very large in the analyses of the maximum lateral diameter *vs.* the maximum dorsoventral diameter (95–99%).

Length of guard vs. depth of pseudoalveolus. – The relationship of these variates is studied in all samples. The correlation coefficients are highly to very highly significant in all samples except one. The correlation coefficient is not significant in the sample from Muleby Å, and this is most likely due to the small number of specimens in this sample. The relationship of the variates may be regarded as isometric.

Length of guard vs. dorsoventral diameter at alveolar end. – The relationship of these variates is studied in all samples. The correlation coefficients are highly to very highly significant in all samples. The relationship of the variates may be regarded as isometric in the samples from Arnager, Muleby Å, Stampe Å, and between Forchhammers Odde and Skidteper. It is allometric to very strongly allometric in the samples from east of Forchhammers Odde, west of Skidteper, Jydegård, and Horsemyre Odde. The reason that some samples show an isometric relationship of the variates is most likely the small number of specimens in these samples and the near lack of juvenile specimens in some of the samples. The growth lines are therefore determined less accurately. In the samples showing an allometric relationship of the variates, adult specimens are stouter than juvenile specimens. Consequently, the means of the Slenderness-Quotient are not calculated in these samples. If the correlation coefficients are perfect (*r*=1.0000) in the samples from east of Forchhammers Odde, Jydegård, and Horsemyre Odde, the Slenderness-Quotient will vary as shown in Tables 50–52. Because of individual variation, the Slenderness-Quotient

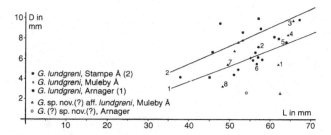

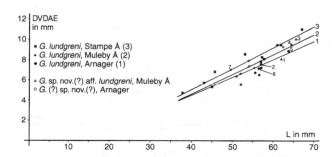

Fig. 51. Scatter plots and regression lines of the length of the guard (L) *vs.* the depth of the pseudoalveolus (D) of *Goniocamax l. lundgreni* from the Arnager Limestone Formation. Regression line 1 is calculated on the basis of material from Arnager and line 2 on the basis of material from Stampe Å. 1, MGUH 1517, lectotype, Stolley (1897, Pl. 3:18); 2, lectotype of *Actinocamax bornholmensis* Stolley (1897, Pl. 4:1); 3, MGUH 1518 (Stolley 1897, Pl. 3:19); 4, MGUH 1519 (Stolley 1897, Pl. 3:20); 5, MGUH 1690 (Ravn 1918, Pl. 2:7); 6, MGUH 7837 (Birkelund 1957, Pl. 1:5); 7, MGUH 1515 (Stolley 1897, Pl. 3:16); 8, MGUH 1516 (Stolley 1897, Pl. 3:17). +=mean values.

Fig. 52. Scatter plots and regression lines of the length of the guard (L) *vs.* the dorsoventral diameter at the alveolar end (DVDAE) of *Goniocamax l. lundgreni* from the Arnager Limestone Formation. Regression line 1 is calculated on the basis of material from Arnager, line 2 on the basis of material from Muleby Å, and line 3 on the basis of material from Stampe Å. +=mean values. For explanation see Fig. 51.

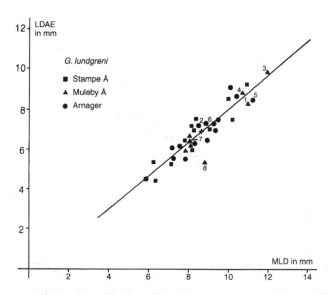

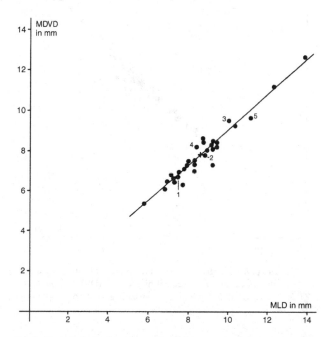

Fig. 54. Scatter plots and regression line of the maximum lateral diameter (MLD) *vs.* the lateral diameter at the alveolar end (LDAE) of *Goniocamax l. lundgreni* from the Arnager Limestone Formation. The regression line is calculated on the basis of material from the Arnager Limestone at Arnager. +=mean values. For explanation see Fig. 51.

Fig. 53. Scatter plot and regression line of the maximum lateral diameter (MLD) *vs.* the maximum dorsoventral diameter (MDVD) of *Goniocamax l. lundgreni* from the Arnager Limestone at Arnager. 1, MGUH 1691 (Ravn 1918, Pl. 2:8); 2, MGUH 7837 (Birkelund 1957, Pl. 1:5); 3, MGUH 7838 (Birkelund 1957, Pl. 1:6); 4, lectotype of *Actinocamax bornholmensis* Stolley (1897, Pl. 4:1); 5, MGUH 1690 (Ravn 1918, Pl. 2:7). +=mean values.

varies from 5.3 (an adult specimen) to 8.3 (a juvenile specimen) in the sample from east of Forchhammers Odde, from 6.0 (an adult specimen) to 9.7 (a juvenile specimen) in the sample from Jydegård, and from 5.7 (an adult specimen) to 7.6 (an adolescent specimen) in the sample from Horsemyre Odde.

Length of guard vs. maximum lateral diameter. – The relationship of these variates is studied in three samples: west

of Arnager, east of Forchhammers Odde, and Jydegård. The correlation coefficients are very highly significant. The relationship of the variates is strongly allometric in the sample from east of Forchhammers Odde, very strongly allometric in the sample from Jydegård, and isometric in the sample from Arnager. The reason that the relationship is isometric in the latter sample may be the small number of specimens; the growth line is determined less accurately.

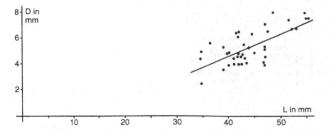

Fig. 55. Scatter plot and regression line of the length of the guard (L) *vs.* the depth of the pseudoalveolus (D) of *Goniocamax l. lundgreni* from the Bavnodde Greensand, east of Forchhammers Odde. +=mean values.

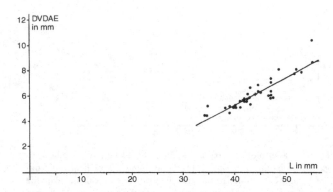

Fig. 56. Scatter plot and regression line of the length of the guard (L) *vs.* the dorsoventral diameter at the alveolar end (DVDAE) of *Goniocamax l. lundgreni* from the Bavnodde Greensand, east of Forchhammers Odde. +=mean values.

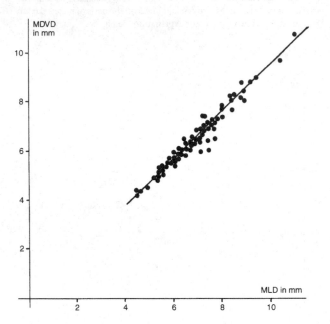

Fig. 57. Scatter plot and regression line of the maximum lateral diameter (MLD) *vs.* the maximum dorsoventral diameter (MDVD) of *Goniocamax l. lundgreni* from the Bavnodde Greensand, east of Forchhammers Odde.

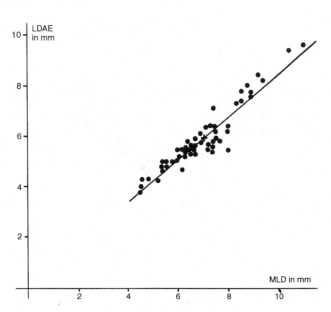

Fig. 58. Scatter plot and regression line of the maximum lateral diameter (MLD) *vs.* the lateral diameter at the alveolar end (LDAE) of *Goniocamax l. lundgreni* from the Bavnodde Greensand, east of Forchhammers Odde. +=mean values.

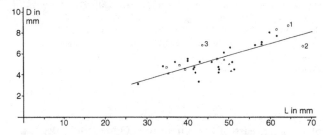

Fig. 59. Scatter plot and regression line of the length of ther guard (L) *vs.* the depth of the pseudoalveolus (D) of *Goniocamax l. lundgreni* from the Bavnodde Greensand at Jydegård. *Actinocamax lundgreni excavata sensu* Birkelund (1957) is plotted as open circles, *Actinocamax westfalicus sensu* Birkelund as open triangles, and *A.* aff. *westfalicus sensu* Birkelund as open squares. 1, MGUH 7840 (Birkelund 1957, Pl. 1:9) (Pl. 2:9, herein), syntype of *G. lundgreni postexcavata* Naidin, 1964; 2, MGUH 7939 (Birkelund 1957, Pl. 1:7) (Pl. 2:8, herein), lectotype of *G. lundgreni postexcavata*; 3, MGUH 7843 (Birkelund 1957, Pl. 2:3) (Pl. 2:11, herein). +=mean values.

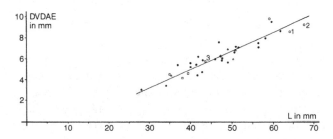

Fig. 60. Scatter plot and regression line of the length of the guard (L) *vs.* the dorsoventral diameter at the alveolar end (DVDAE) of *Goniocamax l. lundgreni* from the Bavnodde Greensand at Jydegård. +=mean values. For explanation see Fig. 59.

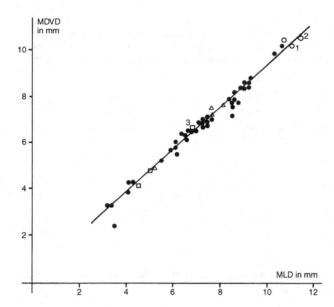

Fig. 61. Scatter plot and regression line of the maximum lateral diameter (MLD) *vs.* the maximum dorsoventral diameter (MDVD) of *Goniocamax l. lundgreni* from the Bavnodde Greensand at Jydegård. For explanation see Fig. 59.

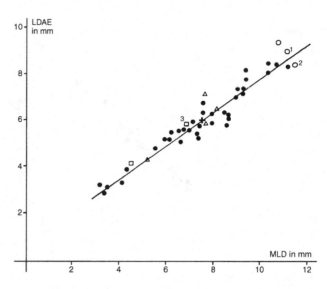

Fig. 62. Scatter plot and regression line of the maximum lateral diameter (MLD) *vs.* the lateral diameter at the alveolar end (LDAE) of *Goniocamax l. lundgreni* from the Bavnodde Greensand at Jydegård. +=mean values. For explanation see Fig. 59.

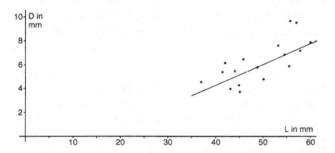

Fig. 63. Scatter plot and regression line of the length of the guard (L) *vs.* the depth of the pseudoalveolus (D) of *Goniocamax l. lundgreni* from the Bavnodde Greensand at Horsemyre Odde. +=mean values.

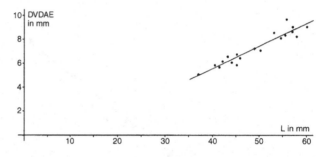

Fig. 64. Scatter plot and regression line of the length of the guard (L) *vs.* the dorsoventral diameter at the alveolar end (DVDAE) of *Goniocamax l. lundgreni* from the Bavnodde Greensand at Horsemyre Odde. +=mean values.

Table 50. Tabulation of the Slenderness-Quotient (SQ) with increasing length of the guard (L) in *Goniocamax l. lundgreni* from the Bavnodde Greensand, east of Forchhammers Odde. Owing to allometry, SQ will decrease when the length of the guard increases. See text for discussion.

L (mm)	SQ
30	10.5
40	7.5
50	6.6
60	6.2

Table 51. Tabulation of the Slenderness-Quotient (SQ) with increasing length of the guard (L) in *Goniocamax l. lundgreni* from the Bavnodde Greensand, Jydegård. Owing to allometry, SQ will decrease when the length of the guard increases. See text for discussion.

L (mm)	SQ
30	8.9
40	7.8
50	7.2
60	6.9
70	6.7

Table 52. Tabulation of the Slendernes-Quotient (SQ) with increasing length of the guard (L) in *Goniocamax l. lundgreni* from the Bavnodde Greensand, Horsemyre Odde. Owing to allometry, SQ will decrease when the length of the guard increases. See text for discussion.

L (mm)	SQ
30	8.0
40	7.1
50	6.7
60	6.4

Maximum lateral diameter vs. maximum dorsoventral diameter. – The relationship of these variates is studied in four samples: west of Arnager, east of Forchhammers Odde, Jydegård, and Horsemyre Odde. The correlation coefficients are very highly significant in all samples. The relationship of the variates is isometric in the samples

from Arnager, east of Forchhammers Odde, and Horsemyre Odde. In these samples the ventral flattening is thus similar in juvenile and adult specimens.

The relationship of the variates is very strongly allometric in the sample from Jydegård. Owing to allometry, adult specimens are more flattened ventrally than juvenile

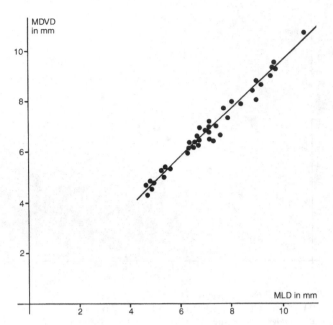

Fig. 65. Scatter plot and regression line of the maximum lateral diameter (MLD) *vs.* the maximum dorsoventral diameter (MDVD) of *Goniocamax l. lundgreni* from the Bavnodde Greensand at Horsemyre Odde.

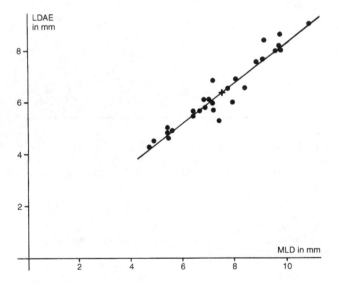

Fig. 66. Scatter plot and regression line of the maximum lateral diameter (MLD) *vs.* the lateral diameter at the alveolar end (LDAE) of *Goniocamax l. lundgreni* from the Bavnodde Greensand at Horsemyre Odde. +=mean values.

specimens. If the correlation coefficient is perfect ($r=1.0000$), the Flattening-Quotient varies as shown in Table 53.

Because of individual variation, the Flattening-Quotient varies from 0.97 (a juvenile specimen) to 1.13 (an adolescent specimen). Because of allometry it is not legal to calculate the mean of the Flattening-Quotient of this

Table 53. Tabulation of the Flattening-Quotient (FQ) with increasing maximum lateral diameter (MLD) in *Goniocamax l. lundgreni* from the Bavnodde Greensand, Jydegård. Owing to allometry, FQ will increase when the maximum lateral diameter increases. See text for discussion.

MLD (mm)	FQ
3	1.00
4	1.02
5	1.03
6	1.04
7	1.05
8	1.05
9	1.06
10	1.06
11	1.06
12	1.07

sample. Nevertheless, we have calculated this mean for comparative purpose.

Maximum lateral diameter vs. lateral diameter at alveolar end. – The relationship of these variates is studied in four samples: west of Arnager, east of Forchhammers Odde, Jydegård, and Horsemyre Odde. The correlation coefficients are very highly significant, and the relationship of the variates is isometric in all samples. The shape of the guard in ventral view is thus similar in juvenile and adult specimens.

Discussion. – The means and observed ranges of the length of the guard, Riedel-Index, Flattening-Quotient, and MLD/LDAE Index of the samples from Arnager, Muleby Å, Stampe Å, east of Forchhammers Odde, Jydegård, and Horsemyre Odde are shown in Fig. 67. We have not plotted the samples from between Forchhammers Odde and Skidteper, and west of Skidteper, because these are small and, by and large, coeval with the sample from east of Forchhammers Odde.

The means of the length of the guard of the three samples from the Arnager Limestone Formation are very closely similar (weighted grand mean is 55 mm), as is the maximum length of the guard (*c.* 65 mm) (Fig. 67). The means of the Flattening-Quotient and MLD/LDAE Index of the three samples are also very closely similar. The weighted grand mean of the MLD/LDAE Index is 1.25, and the weighted grand mean of the Flattening-Quotient is 1.09.

The three samples differ, however, with respect to the depth of the pseudoalveolus and slenderness (Figs. 51–52, 67). The means of the Riedel-Index and Slenderness-Quotient of the samples from Arnager and Stampe Å are compared with the following results. The *F*-test of the variances of the Riedel-Index gives a value of 5.1625, with 8 and 10 degrees of freedom, which is significant ($0.02>P>0.002$). Since the difference is only slightly significant, the means are tested by the *t*-test. This gives a value of 3.5771, with 18 degrees of freedom, which is

Fig. 67. Diagram showing the mean values and observed ranges of the length of the guard, Riedel-Index, Flattening-Quotient, and MLD/LDAE Index of six samples of *Goniocamax l. lundgreni* from the Coniacian and Lower Santonian of Bornholm. 1, Arnager; 2, Muleby Å; 3, Stampe Å (Arnager Limestone Formation); 4, east of Forchhammers Odde; 5, Jydegård; and 6, Horsemyre Odde (Bavnode Greensand Formation).

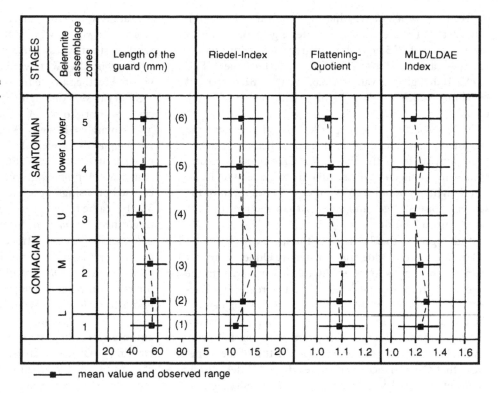

—■— mean value and observed range

highly significant ($0.005 > P > 0.001$). The *F*-test of the variances of the Slenderness-Quotient gives a value of 1.4400, with 9 and 7 degrees of freedom, which is not significant ($P > 0.20$). The *t*-test gives a value of 2.1389, with 16 degrees of freedom, which is significant ($0.05 > P > 0.025$). The specimens from Stampe Å thus have a deeper pseudoalveolus and are stouter than the specimens from Arnager. The specimens from Muleby Å are situated in between the specimens from Arnager and Stampe Å.

The age of the three samples is different. The sample from west of Arnager is lower Lower Coniacian, the one from Muleby Å upper Lower Coniacian, and the one from Stampe Å Middle Coniacian. The samples thus show the following parallel trends in the Coniacian: (1) the pseudoalveolus becomes deeper and (2) the guard becomes stouter.

The four samples from the Bavnodde Greensand exposed east of Forchhammers Odde, between Forchhammers Odde and Skidteper, west of Skidteper (all Upper Coniacian), and Horsemyre Odde (lower Lower Santonian), are very closely similar with respect to the size, slenderness, and ventral flattening of the guard, depth of the pseudoalveolus, and shape in ventral view (Figs. 67–68). The weighted grand means of the four samples are as follows: length of guard: 45 mm; Riedel-Index: 12.3; Flattening-Quotient: 1.05; and MLD/LDAE Index: 1.18. The comparison of the regression lines of the length of the guard *vs.* the dorsoventral diameter at the alveolar end of the samples from east of Forchhammers Odde and

Horsemyre Odde shows that the *F*-test of the variances, and *t*-tests of the slopes and positions do not differ significantly at the 5% level.

The four samples from the Bavnodde Greensand Formation differ from the three samples from the Arnager Limestone Formation in that the guard is smaller, less flattened ventrally, and less lanceolate in ventral view (Fig. 67). The means of the length of the guard, Flattening-Quotient, and MLD/LDAE Index of the samples from Arnager and east of Forchhammers Odde are compared with the following result. The *t*-test of the means of the length of the guard gives a value of 5.9031, with 51 degrees

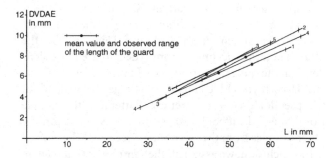

Fig. 68. Regression lines of the length of the guard (L) *vs.* the dorsoventral diameter at the alveolar end (DVDAE) of *Goniocamax l. lundgreni* from the Arnager Limestone and Bavnodde Greensand Formations. 1, Arnager; 2, Stampe Å; 3, east of Forchhammers Odde; 4, Jydegård; and 5, Horsemyre Odde.

of freedom, which is very highly significant ($P<0.001$). The *t*-test of the means of the Flattening-Quotient Index gives a value of 2.5539, with 67 degrees of freedom, which is significant ($0.02>P>0.01$). The *t*-test of the MLD/LDAE Index gives a value of 2.6319, with 81 degrees of freedom, which is very highly significant ($0.01>P>0.005$).

With respect to the depth of pseudoalveolus, the weighted grand mean of the Riedel-Index (12.3) of the four samples from the Bavnodde Greensand is closely similar to the mean of the sample from Muleby Å (12.7). The weighted grand mean is larger than the mean of the sample from Arnager and smaller than the mean of the sample from Stampe Å.

With regard to the slenderness, the specimens from east of Forchhammers Odde and Horsemyre Odde are stouter than the specimens Arnager but are closely similar to the specimens from Stampe Å (Fig. 68). The regression lines of the length of the guard *vs.* the dorsoventral diameter at the alveolar end of the samples from Arnager and east of Forchhammers Odde are compared with the following result. The *F*-test of the variances and *t*-test of the slopes show that the variances and slopes do not differ significantly at the 5% level. The *t*-test of the positions of the regression lines, however, gives a value of 4.3587, with 42 degrees of freedom, which is very highly significant ($P<0.001$). The regression lines of the samples from Stampe Å and east of Forchhammers Odde are also compared. The *F*-test and *t*-tests of the slopes and positions show that the variances, slopes, and positions do not differ significantly at the 5% level.

The sample from the Bavnodde Greensand at Jydegård (basal Lower Santonian) is aberrant in several aspects. As regards the maximum length of the guard, slenderness, and shape in ventral view, it shows close affinity to the samples from the Arnager Limestone Formation, i.e. the guard is large, slender, and markedly flattened ventrally (Figs. 67–68). The mean of the length of the guard is smaller than the means of the samples from the Arnager Limestone Formation, but this is due to the fact that the sample from Jydegård includes more juvenile specimens. The specimens from Jydegård are stouter than the specimens from Arnager, but differ in no significant respect from the specimens from Stampe Å (Fig. 68). On the other hand, the specimens from Jydegård are more slender than the specimens from east of Forchhammers Odde and Horsemyre Odde (Fig. 68). As regards the depth of the pseudoalveolus and ventral flattening, the sample from Jydegård differs in no significant respect from the other samples from the Bavnodde Greensand (Fig. 67).

In conclusion, we assign all the samples to *G. lundgreni lundgreni*, although differences have been revealed. However, these are not considered to be of such taxonomical significance as to establish new species or subspecies. If the sample from Jydegård is not considered, the remaining samples show the following trends through the Coniacian – lower Lower Santonian. The guard becomes smaller, stouter, less flattened ventrally, and less lanceolate in ventral view upwards. These trends, however, are violated by the sample from Jydegård.

Stolley (1897) distinguished three taxa from the Coniacian of Bornholm: *Actinocamax lundgreni* from the 'Glass-marl' at Muleby Å; *A. mammillatus mut. (ant.) bornholmensis* Stolley, lectotype, here designated, the original of Stolley (1897, Pl. 4:1) from the Arnager Limestone; and *A. propinquus mut. (var.) nov.* The latter taxon is based on a single specimen figured by Stolley (1897, Pl. 3:23); it came from the greensand at Stampen. Stolley distinguished *A. lundgreni* from *A. mammillatus* mut. *bornholmensis* by its concentric exfoliation of the anterior end of the guard and distinct longitudinal striation. Ravn (1946) placed *A. mammillatus* mut. *bornholmensis* in synonymy with *A. lundgreni*, because *A. mammillatus* mut. *bornholmensis* also may be exfoliated anteriorly and possess a distinct striation. This view was followed by later authors, including Birkelund (1957), Jeletzky (1958), Naidin (1964b), and Christensen (1973, 1991). As shown above, the specimens from Muleby Å are closely similar to the specimens from west of Arnager. They only differ by being slightly stouter and having a slightly deeper pseudoalveolus. These differences are not regarded to be of such taxonomical value as to distinguish chronological subspecies.

The type specimen of *A. propinquus* mut. (var.) nov. is apparently lost, because it is not in the collections of the Geologisch-Paläontologisches Institut, Greifswald, as mentioned by Stolley (1897). Stolley (1896, 1897) distinguished this taxon by its well-developed vascular imprints, ventral flattening, and subtriangular cross-section of the anterior end. He also noted that this taxon is smaller and has a more shallow pseudoalveolus than *Belemnitella propinqua*. The specimens from Stampe Å are closely similar to the specimens from west of Arnager. They only differ by being slightly stouter and having a slightly deeper pseudoalveolus. The differences are not considered as being of such taxonomical value as to refer the specimens to a new species or subspecies. Since the original of *A. propinquus* mut. (var.) nov. is very closely comparable to other specimens of *G. lundgreni lundgreni* from Stampe Å (cf. Pl. 2:7), *A. propinquus* mut. (var.) nov. is placed in synonymy with *G. lundgreni lundgreni*.

Birkelund (1957) recorded *Goniocamax lundgreni excavata* from Jydegård; *Gonioteuthis westfalica* from Jydegård, Risenholm, Horsemyre Odde, and the district between Forchhammers Odde and Horsemyre Odde; and *Gonioteuthis* aff. *westfalica* from Jydegård, Horsemyre Odde, and the district between Forchhammers Odde and Horsemyre Odde.

G. lundgreni excavata was distinguished from the nominotypical subspecies by its larger guard and fairly distinct vascular imprints. *G. lundgreni excavata sensu* Birkelund has indeed more prominent vascular markings than the nominotypical subspecies, but the depth of the pseudoalveolus is very closely similar to the nominotypical species (see above). It is here placed in synonymy with *G. lundgreni lundgreni*, in spite of its more prominent vascular markings.

We have analyzed in detail *G.* aff. *westfalica sensu* Birkelund, *G. westfalica sensu* Birkelund, and *G. lundgreni excavata sensu* Birkelund from Jydegård; these are plotted in Figs. 59–62. They only differ by their size and thus represent different growth stages of *G. lundgreni lundgreni*. Birkelund referred juvenile specimens of *G. lundgreni lundgreni* to *G.* aff. *westfalica*, adolescent specimens to *G. westfalica*, and adult specimens to *G. lundgreni excavata*. The specimens of *G. westfalica sensu* Birkelund and *G.* aff. *westfalica sensu* Birkelund from Horsemyre Odde, and the district between Forchhammers Odde and Horsemyre Odde are also placed in synonymy with *G. lundgreni lundgreni*.

Naidin (1964b) recorded *G. lundgreni lundgreni* and *G. lundgreni excavata* from the Upper Coniacian of the Russian Platform, and they were considered as geographical subspecies. He noted that the former subspecies is distributed in the western part and the latter subspecies in the eastern part of the Russian Platform. According to Naidin, the two subspecies are closely comparable and distinguished by the structure and cross-section of the alveolar end. *G. lundgreni lundgreni* has a well preserved alveolar end, the cross-section of which is egg-shaped, and *G. lundgreni excavata* is usually exfoliated anteriorly and the cross-section is subtriangular. The specimens of *G. lundgreni lundgreni* from Bornholm before us show a wide variation with respect to these characters, which are not considered of taxonomical importance. We therefore place *G. lundgreni excavata* in synonymy with *G. lundgreni lundgreni*. The specimens of *G. lundgreni lundgreni* from Russia, however, attain a larger size of the guard than the specimens from Bornholm. The maximum length of the guard of the Bornholm specimens is ca. 70 mm, whereas the maximum length of the guard of the specimens from Russia is up to 85 mm.

Naidin (1964b) established *Gonioteuthis (Goniocamax) lundgreni postexcavata* and referred the two specimens, figured as *Actinocamax lundgreni excavata* by Birkelund (1957, Pl. 1:7–8), to his new taxon. He tentatively also included two specimens from the Volga area, figured as *B. propinqua* by Jeletzky (1949, Figs. 1–2), in his new subspecies. The specimens from Bornholm and the specimens from the Volga area are not conspecific (see below). Naidin did not designate a type for *G. (G.) lundgreni postexcavata*, and MGUH 7839, the original of Birkelund (1957, Pl. 1:7), is here designated as lectotype for this subspecies. *G. (G.) lundgreni postexcavata*, as interpreted with respect to its lectotype and syntype, differs in no significant

respect from *G. l. lundgreni* (see above) and is therefore placed in synonymy with *G. l. lundgreni*.

Naidin (1974) emended the diagnosis of *G. (G.) lundgreni postexcavata*, but oddly enough he noted that the two specimens from Bornholm and the two from the Volga area probably are similar to *G. (G.) lundgreni postexcavata*. The specimens figured as *G. (G.) lundgreni postexcavata* by Naidin (1974, Pl. 73:10–11) are here assigned to *B. schmidi* sp. nov. with a query (see below).

G. lundgreni lundgreni is smaller than *G. lundgreni uilicus* (Koltypin) (see Naidin 1964b), and has a more shallow pseudoalveolus and is stouter than *G. mirabilis*. The affinity to *B. schmidi* is discussed below.

G. esseniensis (Christensen, 1982), from the lower Lower Coniacian of the Münster Basin, is closely allied to *G. lundgreni lundgreni* with respect to the shape of the guard, slenderness, and ventral flattening, but it is larger than the specimens from Bornholm and has a more shallow pseudoalveolus.

Distribution. – *G. lundgreni lundgreni* is mainly distributed in Baltoscandia and the Central Russian Subprovince. It is also recorded from Germany and England (Christensen 1991). It occurs from the Coniacian to the middle Lower Santonian.

Goniocamax sp. nov.(?) aff. *lundgreni* (Stolley, 1897)

Pl. 2:6

Synonymy. – ☐ pars 1946 *Actinocamax lundgreni* Stolley – Ravn, pp. 30–31 (only specimen no. 8). ☐ pars 1957 *Actinocamax lundgreni lundgreni* Stolley – Birkelund, pp. 16–17.

Material. – MGUH 23730, 'Glass-marl' at Muleby Å, loc. 5 of Ravn (1946).

Dimensions. – See Table 54.

Description. – Guard medium-sized and flattened ventrally, lanceolate in ventral view, with the largest diameter situated in the middle part, and subcylindrical in lateral view; anterior end with a very shallow pseudoalveolus; apex with indistinctly separated mucro, displaced towards dorsal side; anterior end subtriangular and compressed laterally; dorsolateral depressions and double furrows, and single lateral furrows present; otherwise smooth guard.

Discussion. – The specimen is very closely comparable to *G. lundgreni* but differs by its very shallow pseudoalveolus. We are doubtful whether the specimen is but an extreme variant of *G. lundgreni* or represents a new species. Since only one specimen is available, it is referred to as *G.* sp.nov.(?) aff. *lundgreni*. It is plotted in Figs. 51–52.

Table 54. Dimensions of *Goniocamax* sp.nov.(?) aff. *lundgreni*, 'Glass-marl' of the Arnager Limestone Formation, Muleby Å, loc. 5. Measurements in millimetres.

Character	MGUH 23730
L	62.4
D	2.6
DVDAE	9.3
LDAE	8.4
MLD	11.5
MDVD	9.8
RQ	24.0
RI	4.2
SQ	6.7
L/MLD	5.4
FQ	1.17
MLD/LDAE	1.37
DVDAE/LDAE	1.11

Table 55. Dimensions of *Goniocamax*(?) sp.nov.(?), bottom conglomerate of the Arnager Limestone, west of Arnager. *=estimated. Measurements in millimetres.

Character	MGUH 23731
L	54*
D	2.7
DVDAE	7.4
LDAE	6.8
MLD	8.5
MDVD	7.7
RQ	20.0
RI	5.0
SQ	7.3
L/MLD	6.7
FQ	1.10
MLD/LDAE	1.25
DVDAE/LDAE	1.09

Table 56. Dimensions of *Goniocamax birkelundae* sp.nov. from the Bavnodde Greensand. GPIK 3909, holotype, Jydegård; MGUH 23732, paratype, Risenholm; MGUH 23733, paratype, east of Forchhammers Odde. Measurements in millimetres.

Character	GPIK 3909	MGUH 23732	MGUH 23733
L	49.0	51.5	50.7
D	6.0	7.7	5.5
DVDAE	7.3	8.5	7.6
LDAE	7.0	8.2	7.0
MLD	8.3	9.5	8.7
MDVD	7.5	8.4	7.5
RQ	8.2	6.7	9.2
RI	12.2	15.0	10.9
SQ	6.7	6.1	6.7
L/MLD	5.9	5.4	5.8
FQ	1.11	1.13	1.16
MLD/LDAE	1.19	1.16	1.24
DVDAE/LDAE	1.04	1.04	1.09

Ravn (1946) discussed this specimen, which was assigned to *A. lundgreni*. Birkelund (1957, pp. 16–17) most likely also discussed this specimen, which she referred to *A. lundgreni lundgreni*. However, she erroneously mentioned that it came from the Arnager Limestone at Arnager.

Distribution. – *Goniocamax* sp.nov.(?) aff. *lundgreni* came from the the 'Glass-marl', loc. 5 at Muleby Å, where it occurs together with *G. lundgreni*. It is upper Lower Coniacian.

Goniocamax(?) sp.nov.(?)

Pl. 1:14

Material. – MGUH 23731, bottom conglomerate of the Arnager Limestone, west of Arnager.

Dimensions. – See Table 55.

Description. – Guard not complete, apical end missing; preserved length 46.5 mm; estimated total length 54 mm; medium-sized, ventrally flattened guard, lanceolate in ventral view, with the largest diameter in the middle part, and subcylindrical in lateral view; flat, oblique anterior end, with a central pit; dorsal side more incised than ventral side; cross section of anterior end subtriangular; dorsolateral depressions and double furrows, and single lateral furrows present; otherwise smooth guard.

Discussion. – This taxon differs from *Goniocamax lundgreni* by the actinocamacoid structure of the anterior end, which is closely similar to that seen in some specimens of *Gonioteuthis westfalica*. However, *Goniocamax*(?) sp. nov.(?) occurs together with *G. lundgreni* and is older than the earliest member of the evolutionary lineage of

Gonioteuthis, G. praewestfalica. We are doubtful whether the specimen is but a very extreme variant of *G. lundgreni* or should be considered an ancestor of the *Gonioteuthis* lineage. It is referred to as *Goniocamax*(?) sp.nov.(?) and is plotted in Figs. 51–52.

Distribution. – *Goniocamax*(?) sp.nov.(?) occurs in the bottom conglomerate of the Arnager Limestone, west of Arnager, which has also yielded *Goniocamax lundgreni*. It is lower Lower Coniacian.

Goniocamax birkelundae sp.nov.

Pl. 3:2–4

Synonymy. – □1957 *Actinocamax* sp., transitional form between *A. westfalicus* and *A. granulatus* – Birkelund, p. 29.

Types. – Holotype is GPIK 3909 from the Bavnodde Greensand, Jydegård. Paratypes are also from the Bavnodde Greensand and include GM 1995.4-5 from Jydegård; MGUH 23732, GM 1995.6-7 from Risenholm; and MGUH 23733 and GPIK 3910-4 from east of Forchhammers Odde. The holotype is basal Lower Santonian, and the paratypes are Upper Coniacian – basal Lower Santonian.

Derivation of name. – The species is named in honour of the late Prof. Tove Birkelund, who monographed the Upper Cretaceous belemnites of Denmark in 1957 and was the first to recognize this form.

Dimensions. – See Table 56.

Diagnosis. – Guard small with a shallow pseudoalveolus, lanceolate in ventral view, subcylindrical in lateral view, and markedly flattened ventrally; surface markings, including dorsolateral longitudinal depression and double furrows, vascular imprints and longitudinal striae, very prominent; internal characters unknown.

Description. – Guard small and markedly flattened ventrally (FQ 1.06–1.17, mean value 1.12), lanceolate in ventral view, with the largest diameter in the middle part or posterior third, and subcylindrical or slightly lanceolate in lateral view; anterior end with a shallow pseudoalveolus (Riedel-Index 11–17, mean value ca. 14; Riedel-Quotient 6–9, mean value ca. 7); apex with indistinctly separated mucro, slightly displaced towards dorsal side; Slenderness-Quotient ca. 6–7; cross-section of anterior end subtriangular and compressed laterally; wall of pseudoalveolus curved or almost straight, with a single or a few conellae; pseudoalveolus often with dorsal and ventral notches; short ventral fissure may be present; dorsolateral longitudinal depressions and double furrows, and single lateral furrows well developed; vascular markings and longitudinal striae prominent; internal characters unknown.

Biometry. – The types were analyzed by univariate statistical methods (Table 57).

Discussion. – The internal characters of *G. birkelundae* are unknown, because none of the guards have been split. *G. birkelundae* is closely allied to *G. lundgreni*, but is more flattened ventrally, has more prominent vascular markings and longitudinal striae, and is smaller. *G. birkelundae* is also closely allied to *G. striatus*, but has a more shallow pseudoalveolus, is more flattened ventrally, and is more lanceolate in ventral view. *G. birkelundae* is most likely the direct ancestor of *G. striatus* sp.nov. *G. birkelundae* differs from *Belemnitella schmidi* sp.nov. by its smaller guard, more shallow pseudoalveolus, more prominent vascular markings, and more prominent longitudinal striae. *G. birkelundae* is smaller and stouter and has a more shallow pseudoalveolus than *G. mirabilis*.

Distribution. – *G. birkelundae* is very rare and only recorded from Bornholm, where it occurs in the Bavnodde Greensand exposed east of Forchhammers Odde (Upper Coniacian), Risenholm (uppermost Coniacian – basal Lower Santonian), and Jydegård (basal Lower Santonian)

Goniocamax striatus sp.nov.

Pl. 3:5–7

Types. – The types are from the Bavnodde Greensand and are upper Lower Santonian. Holotype is GPIK 3915, west of Bavnodde, between beds 2 and 4. Paratypes include GM 1995.3, west of Bavnodde, between beds 5 and 6; MGUH 23734, west of Bavnodde, bed 5; GPIK 3916-7,

Table 57. Univariate analysis of *Goniocamax birkelundae* sp.nov. from the Bavnodde Greensand. Measurements in millimetres.

Character	N	$\bar{X}$	SD	CV	OR
L	8	48.6	3.0	6.2	44.5–51.5
D	10	6.7	1.1	16.2	5.4–8.5
DVDAE	10	7.3	0.6	8.3	6.6–8.5
LDAE	9	6.9	0.7	9.8	6.0–8.2
MLD	11	8.3	1.0	12.1	6.2–9.5
MDVD	11	7.4	0.9	11.6	5.8–8.8
RQ	8	7.3	1.1	15.1	6.1–9.2
RI	8	13.9	2.0	14.1	10.9–16.5
SQ	8	6.9	0.6	8.3	6.1–7.2
L/MLD	8	5.7	0.3	5.6	5.4–6.4
FQ	11	1.12	0.03	3.0	1.06–1.17
MLD/LDAE	9	1.22	0.07	6.0	1.09–1.33
DVDAE/LDAE	10	1.05	0.02	2.3	1.00–1.09

Table 58. Dimensions of *Goniocamax striatus* sp.nov. from the Bavnodde Greensand. GPIK 3915, holotype, east of Bavnodde, between beds 2 and 4; MGUH 23734, paratype, west of Bavnodde, bed 5; MGUH 23735, paratype, east of Bavnodde, bed 21. * = estimated. Measurements in millimetres.

Character	GPIK 3915	MGUH 23734	MGUH 23735
L	48.0	51*	39.4
D	9.7	10.2	7.9
DVDAE	7.6	7.1	5.5
LDAE	7.6	7.1	5.3
MLD	8.9	7.7	6.1
MDVD	7.9	7.3	5.4
LVF	1.9	–	–
SD	–	*c.* 4.7	–
FA	–	*c.*45	–
RQ	5.0	5.0	7.1
RI	20.2	20.0	20.1
SQ	6.3	7.2	7.2
L/MLD	5.4	6.6	6.5
FQ	1.13	1.05	1.13
MLD/LDAE	1.17	1.08	1.15
DVDAE/LDAE	1.00	1.00	1.04

west of Bavnodde, between beds 2 and 3; GPIK 3918, west of Bavnodde, between beds 1 and 2; and MGUH 23735, east of Bavnodde, bed 21.

Derivation of name. – Latin *striatus*, striate.

Dimensions. – See Table 58.

Diagnosis. – Guard small with a deep pseudoalveolus, lanceolate in ventral view, subcylindrical in lateral view, and markedly flattened ventrally; surface markings, including dorsolateral longitudinal depressions and double furrows, vascular imprints and longitudinal striae, very prominent; Schatzky distance large; fissure angle large.

Description. – Guard small and markedly flattened ventrally (Flattening-Quotient 1.05–1.13, mean value 1.09), lanceolate in ventral view, with the largest diameter in the middle part or posterior third, and subcylindrical or

Table 59. Univariate analysis of *Goniocamax striatus* sp.nov. from the Bavnodde Greensand. Measurements in millimetres.

Character	N	X̄	SD	CV	OR
L	7	46.2	3.5	7.7	39.4–51.0
D	7	9.3	1.0	10.2	7.9–10.5
DVDAE	7	6.6	0.8	11.5	5.7–7.6
LDAE	7	6.5	0.9	13.1	5.4–7.6
MLD	7	7.2	1.0	13.3	6.1–8.9
MDVD	7	6.7	0.9	13.0	5.4–7.9
RQ	7	5.0	0.3	5.6	4.5–5.4
RI	7	20.1	1.2	5.8	18.5–22.3
SQ	7	7.0	0.5	7.0	6.3–7.9
L/MLD	7	6.4	0.6	8.9	5.4–7.3
FQ	7	1.09	0.04	3.4	1.05–1.13
MLD/LDAE	7	1.12	0.04	3.6	1.07–1.17
DVDAE/LDAE	7	1.03	0.03	2.5	1.00–1.06

slightly lanceolate in lateral view; anterior end with a deep pseudoalveolus (Riedel-Index 18.5–22.5, mean value 20; Riedel-Quotient 4.5–5.5, mean value 5); apex with indistinctly separated mucro, slightly displaced towards dorsal side; Slendernes-Quotient 6–8; cross-section of the anterior end subtriangular and compressed laterally; wall of the pseudoalveolus curved or almost straight, with a single or a few conellae; pseudoalveolus often with dorsal and ventral notches; short ventral fissure may be present; dorsolateral longitudinal depressions and double furrows, and single lateral furrows well developed; vascular markings and longitudinal striae very distinct; Schatzky distance large, ca. 5 mm, and fissure angle large, ca. 45°.

Biometry. – The types were analyzed by univariate statistical methods (Table 59).

Discussion. – The internal characters of *G. striatus* are known only in one specimen, MGUH 23734 (Table 58). The species is closely allied to *G. lundgreni*, but is distinguished by its smaller guard, deeper pseudoalveolus, more prominent vascular markings and longitudinal striae, and possibly larger Schatzky distance. *G. striatus* is most likely the lineal descendant of *G. birkelundae* sp.nov. (see above). *G. striatus* differs from *B. schmidi* sp.nov. by its smaller size and more prominent surface markings. *G. striatus* differs from *G. mirabilis* by its smaller and stouter guard, which is less lanceolate in ventral view.

Distribution. – *G. striatus* is very rare and recorded only from Bornholm, where it occurs in the Bavnodde Greensand exposed east and west of Bavnodde, which is upper Lower Santonian.

Genus *Belemnitella* d'Orbigny, 1840

Type species. – *Belemnites mucronatus* Schlotheim, 1813; ICZN Opinion 1328 (1985); name no. 2979.

Diagnosis. – See Christensen (1975a).

Discussion. – *B. propinqua* was regarded to be the earliest representative of the genus by Jeletzky (1949, 1955), Naidin (1964a, 1964b, 1974), and Christensen (1971, 1974, 1991, 1993a), although other authors, including Birkelund (1957) and Glazunova (1972), placed this species in the genus *Actinocamax*. In the present paper, we place *B. propinqua* as well as *B. schmidi* sp.nov. in *Belemnitella*, because *B. schmidi* sp.nov. is considered as the direct ancestor of *B. propinqua*.

Distribution. – *Belemnitella* is widely distributed in the North European Province and has been recorded also from the northern part of the Tethyan Realm and the North American Province. The genus occurs from the basal Lower Santonian to the uppermost Maastrichtian.

Belemnitella schmidi sp.nov.

Pl. 3:8–10

Synonymy. – ☐1949 *Belemnitella propinqua* (Moberg) – Jeletzky, p. 416, Figs. 1–2 (*non* Figs. 3–4). ☐?1974 *Gonioteuthis* (*Goniocamax*) *lundgreni postexcavata* – Naidin, p. 211, Pl. 73:10–11.

Types. – Holotype is GPIK 3919, from the Bavnodde Greensand at Jydegård, basal Santonian; paratypes are two unregistered specimens, GSC, *ex* J.A. Jeletzky Collection, from the Lower Santonian, lower(?) part of *Inoceramus pachti* Zone *sensu rossico*, Volga area. The paratypes were figured as *Belemnitella propinqua* by Jeletzky (1949, Figs. 1–2). GPIK 3920 from Jydegård, a ventrally flattened apical fragment of a large belemnite probably belongs to *B. schmidi*

Derivation of name. – The species is named in honour of Professor Friedrich Schmid.

Dimensions. – See Table 60.

Diagnosis. – Guard large with a shallow pseudoalveolus, lanceolate in ventral view, subcylindrical in lateral view, and markedly flattened ventrally; surface markings, including dorsolateral longitudinal depressions and double furrows, vascular imprints, and striae, well developed; Schatzky distance small; fissure angle large.

Description. – Guard large and slender, markedly lanceolate in ventral view, with largest diameter in posterior third, and subcylindrical in lateral view; guard flattened ventrally over its entire length; apex with mucro displaced towards dorsal side; anterior end with shallow pseudoalveolus (Riedel-Index 15–19); cross-section of anterior end subtriangular or pointed oval; wall of pseudoalveolus straight, with conellae; Schatzky distance small and fissure angle large; bottom of ventral fissure s-shaped; surface

Table 60. Dimensions of *Belemnitella schmidi* sp. nov. GPIK 3919, holotype, Bavnodde Greensand, Jydegård; 1, paratype, *B. propinqua* of Jeletzky (1949, Fig. 1) (Pl. 3:8 herein); 2, paratype, *B. propinqua* of Jeletzky (1949, Fig. 2) (Pl. 3:9 herein). *=estimated. Measurements in millimetres.

Character	1	2	GPIK 3919
L	66*	80*	80*
D	12.3*	12*	13.5*
LAP	53*	68*	66.5*
DVDP	10.1	12.8	13.1
LDP	10.2	12.5	12.3
MLD	12.1	14.7	15.9
MDVD	10.6	13.0	13.5
DVDAE	–	11.5	11.5
LDAE	–	11.1	11*
LVF	–	6.8	–
SD	2.3	–	–
FA	46*	–	–
RI	18.6	15.0	16.7
RQ	5.4	6.7	5.9
SQ	–	7.0	7.0
BI	5.3	5.3	5.1
L/MLD	5.5	5.4	5.0
FQ	1.14	1.13	1.18
MLD/LDP	1.19	1.18	1.29

markings, including dorsolateral longitudinal depressions and double furrows, vascular imprints, and striae, well developed.

Discussion. – *B. schmidi* is a transition form possessing characters in common with the genera *Goniocamax* and *Belemnitella*. Goniocamacoid characters include the presence of a pseudoalveolus and the ventrally flattened guard, which is markedly lanceolate in ventral view. It is placed in the genus *Belemnitella*, because it is closely allied to *B. p. propinqua*. It differs from *G. l. lundgreni* by its larger guard and deeper pseudoalveolus. It is closely similar to goniocamacoid specimens *B. p. propinqua* (see below) with respect to size and shape of the guard, in addition to surface markings, but has a more shallow pseudoalveolus. Moreover, *B. schmidi* has a smaller Schatzky distance and a larger fissure angle and is slightly more slender than *B. p. propinqua*. It is worthy of note, however, that the internal characters of *B. schmidi* are known only in one specimen (Table 60). *B. schmidi* is larger and stouter, has a more shallow pseudoalveolus, and is less lanceolate in ventral view than *G. mirabilis*. *B. schmidi* is most likely the direct ancestor of *B. p. propinqua*.

The two specimens figured as *B. propinqua* by Jeletzky (1949, Figs. 1–2) were erroneously placed in synonymy with *Goniocamax lundgreni* by Christensen (1991, p. 734).

Distribution. – *B. schmidi* occurs in the Bavnodde Greensand Formation at Jydegård, which is basal Lower Santonian. It is also recorded from the Lower Santonian *Inoce-*

ramus cardissoides or *I. pachti* Zones *sensu rossico*, Volga area (Jeletzky 1949; Naidin 1974).

Belemnitella propinqua propinqua (Moberg, 1885)

Pl. 1:15–16; Pl. 3:11; Figs. 16B, 69A–D

Synonymy. – □1885 *Actinocamax propinquus* n.sp. – Moberg, p. 53, Pl. 5:25. □1885 *Actinocamax propinquus* n.sp.? – Moberg, p. 53, Pl. 6:22. □1897 *Belemnitella mucronata* Schlotheim *mut. (ant.)* – Stolley, p. 296. □ *non* 1912 *Actinocamax propinquus* Moberg – Arkhangelsky, p. 85, Pl. 10:14–15, 23–27, 34–36. □1921 *Actinocamax propinquus* Moberg – Ravn, p. 38, Pl. 3:2. □ *non* 1948 *Belemnitella propinqua* (Moberg) – Jeletzky, p. 594. □1948 *Belemnitella* ex gr. *mirabilis* Arkhangelsky – Jeletzky, p. 594. □ *non* 1949 *Belemnitella propinqua* (Moberg) – Jeletzky, p. 416, Figs. 1, 2, 4. □1949 *Belemnitella propinqua* (Moberg) – Jeletzky, p. 417, Fig. 3. □1949 *Belemnitella* ex gr. *mirabilis* Arkhangelsky – Jeletzky, p. 422. □1955 *Belemnitella* ex gr. *mirabilis* Arkhangelsky – Jeletzky, p. 481, Pl. 58:5. □ *pars* 1957 *Actinocamax propinquus propinquus* Moberg – Birkelund, p. 20. □1957 *Actinocamax propinquus* Moberg *ravni* n. subsp. – Birkelund, p. 21, Pl. 2:5. □ *non* 1958 *Belemnitella propinqua* (Moberg) – Jeletzky, p. 24, 28. □1958 *Belemnitella* ex gr. *mirabilis* Arkhangelsky – Jeletzky, p. 28. □ *non* 1958 *Actinocamax propinquus* (Moberg) – Nikitin, p. 12, Pl. 1:9–15; Pl. 3:2. □1958 *Belemnitella rylskiana* sp. n. – Nikitin, p. 14, Pl. 3:1–6, 8. □ *non* 1962 '*Belemnitella*' *propinqua* (Moberg) – Kongiel, p. 127. □1962 *Gonioteuthis jeletzkyi* n.sp. – Kongiel, p. 127, footnote. □1964a *Belemnitella propinqua propinqua* (Moberg) – Naidin, p. 85. □ *non* 1964b *Belemnitella propinqua propinqua* (Moberg) – Naidin, p. 167. □1964b *Belemnitella propinqua rylskiana* Nikitin – Naidin, p. 167. □ *non* 1972 *Actinocamax (Actinocamax) propinquus propinquus* Moberg – Glazunova, p. 106, Pl. 45:1–5. □1973 *Belemnitella propinqua propinqua* (Moberg) – Christensen, p. 117, Pl. 9:1–5; Pl. 10:1–3. □ *non* 1974 *Belemnitella propinqua propinqua* (Moberg) – Naidin, p. 213, Pl. 73:12–13. □1974 *Belemnitella propinqua rylskiana* Nikitin – Naidin, p. 213, Pl. 73:14–15. □1991 *Belemnitella propinqua propinqua* (Moberg) – Christensen, p. 736, Pl. 4:13–18; Pl. 5:11–14. □1993a *Belemnitella propinqua* (Moberg) – Christensen, p. 46, fig. 6C.

Type. – Holotype, by monotypy, SGU Type 3906, the original of Moberg (1885, Pl. 5:25), from Eriksdal, Scania, southern Sweden; upper Middle Santonian *Gonioteuthis westfalicagranulata* Zone (Christensen 1986). It was refigured by Christensen (1971, Pl. 1:1) and is figured here as Pl. 1:15.

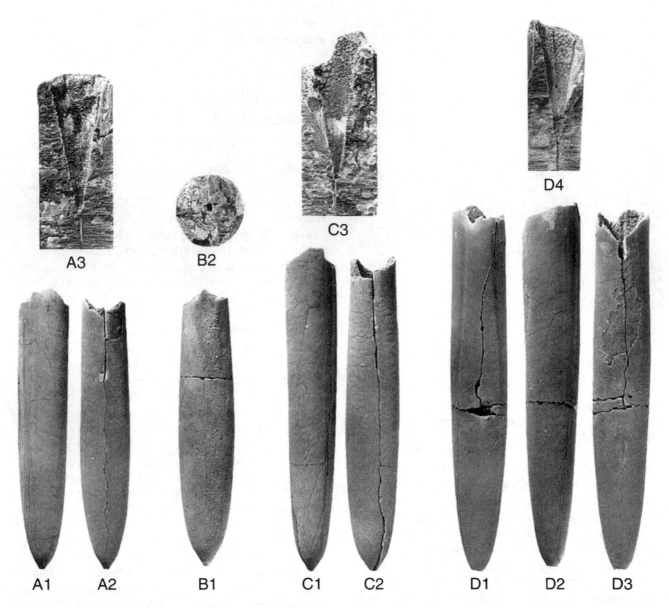

Fig. 69. □A–D. *Belemnitella p. propinqua* (Moberg). All specimens are coated with ammomium chloride and are natural size unless otherwise stated. □A. MGUH 23736, Bavnodde Greensand, east of Bavnodde, between beds 20 and 21. 1, lateral view; 2, ventral view; 3, view of the split anterior end showing internal characters, ×1.5. It is subcylindrical in ventral view (MLD/LDP=1.04). The Riedel-Quotient is 2.7, the Birkelund Index is 3.5, and the Schatzky distance is 6.1 mm. □B. MGUH 23737, Bavnodde Greensand, west of Bavnodde, between beds 2 and 6. 1, ventral view; 2, view of the anterior end, ×1.5. Slender specimen which is lanceolate in ventral view. The Birkelund Index is 5.5, and MLD/LDP Index is 1.2. It has a fracture immediately anterior to the protoconch, probably caused by the bite of an unknown animal. □C. MGUH 23738, Bavnodde Greensand, west of Bavnodde, between beds 2 and 3. 1, lateral view; 2, ventral view; 3, view of the split anterior end showing internal characters, ×1.5. It is subcylindrical in ventral view (MLD/LDP=1.02), and the Birkelund Index is 4.2. □D. GSC 99475, Zone of *Inoceramus* ex gr. *cardissoides sensu rossico*, probably lower part of the Santonian, northeast of the town of Rylsk, Russian Platform. 1, dorsal view; 2, lateral view; 3, ventral view; 4, view of the split anterior end. It was figured as *Belemnitella* ex gr. *B. mirabilis* Arkhangelsky (?=*Belemnitella mucronata* mut. *anterior* Stolley) by Jeletzky (1955, Pl. 58:5).

Material. – Twenty-four specimens from the Bavnodde Greensand Formation: west of Bavnodde (12 specimens), east of Bavnodde (three specimens), Bavnodde (three specimens), Forchhammers Klint (one specimen), Forchhammers Odde (three specimens), and Horsemyre Odde (two specimens). These specimens include MGUH 1807, the holotype of *A. propinquus ravni* Birkelund, 1957, figured by Ravn (1921, Pl. 3:2); MGUH 7845, figured by Birkelund (1957, Pl. 2:5); and MGUH 11611-3, figured by Christensen (1971, Pl. 2:2; Pl. 3:2–3). Precisely located specimens came from west of Bavnodde, between beds 2 and 6, between beds 2 and 3, between beds 3 and 4; and east of Bavnodde, between beds 20 and 21, bed 27.

In addition to the specimens from Bornholm, 96 specimens from southern Sweden and one specimen from the Russian Platform are also included in the present study. These include the holotype; SGU unregistered, one specimen (Pl. 1:16 herein); RM Mo 151004 (Christensen 1971, Pl. 2:3), all from Eriksdal; RM Mo 8348, 'westfalicus-conglomerate' loc. CV 1, Rödmölla–Tosterup (Christensen 1986); SGU Type 5215 from the Kullemölla boring, level 248 m (Christensen 1986, Pl. 3:2); SGU unregistered, one specimen from Gräsryd (Christensen 1993a, Fig. 6C), 90 specimens from Särdal, including MGUH 12730-7 (Christensen 1973, Pl. 9:1–5; Pl. 10:1–3); and GSC 99475 from northeast of the town of Rylsk, Russia (Jeletzky 1955, Pl. 58:5) (Fig. 69D herein).

Dimensions. – See Table 61.

Description. – Guard very large and slender, markedly lanceolate or subcylindrical in ventral view, with largest diameter in middle part or posterior third of guard, and subcylindrical or high cone-shaped in lateral view; relationship between length from apex to protoconch and dorsoventral diameter at protoconch allometric, adult specimens more slender than juvenile specimens; guard flattened or markedly flattened ventrally along its entire length; apical end with indistinctly separated mucro, displaced dorsally; anterior end with pseudoalveolus or alveolus; cross-section of pseudoalveolus subtriangular or pointed oval; cross section of alveolus round; Riedel-Quotient 2.7–4.3, usually ca. 3–4, mean values 3.3–3.4; wall of pseudoalveolus straight or slightly convex, with conellae; conellae small in posterior part of pseudoalveolus and larger in anterior part; Schatzky distance and fissure angle small; shape of bottom of ventral fissure generally straight but also convexly and concavely curved, or s-shaped; vascular imprints present and most prominent around ventral fissure; dorsolateral longitudinal depressions and double furrows distinct; double furrows extend almost to apex; entire surface of guard with longitudinal striae, typically more distinct than vascular imprints; angle between dorsolateral double furrows and main vascular markings apically less than 30°.

Biometry. – A sample from the Bavnodde Greensand was analyzed by univariate and bivariate methods. The specimens are from east and west of Bavnodde, Horsemyre Odde, Forchhammers Odde, and Forchhammers Klint.

Univariate analysis. – The results of the univariate analysis are shown in Table 62.

Bivariate analysis. – The following regression analyses are made: (1) the length from the apex to the protoconch *vs.* the dorsoventral diameter at the protoconch, and (2) the maximum lateral diameter *vs.* the lateral diameter at the protoconch. The equations of the regression lines are given in Table 63, and the scatter plots are shown in Figs. 70–71.

In the first analysis, the correlation coefficient is significant and the relationship of the variates is highly allometric. Since the relationship is allometric, it is not legal to calculate the mean Birkelund Index. In the second analysis, the correlation coefficient is very highly significant, and the relationship of the variates is isometric. If the correlation coefficient is perfect ($r = 1.0000$), the ratio

Table 61. Dimensions of *Belemnitella p. propinqua.* SGU Type 3906, holotype, Eriksdal, upper Middle Santonian; 1, *Actinocamax propinquus?* Moberg (1885, Pl. 6:22) = *Belemnitella mucronata* mut. *anterior* Stolley (1897), Pl. 1:16 herein; RM Mo 151004, Eriksdal, upper Middle Santonian; GSC 99475, Fig. 69D herein. *=estimated; **=without mucro. Measurements in millimetres.

Character	SGU Type 3906	1	RM Mo 151004	GSC 99475
L	78.0**	–	–	91*
D	22.3	–	–	33.5*
LAP	55.7	53.0	–	61*
DVDP	13.5	11.7	13.5*	14*
LDP	13.7	11.8	14.0*	14.2
MLD	15.0	12.6	–	14.1
MDVD	12.9	11.7	–	14.0
DVAE	12.6	–	–	–
LDAE	11.7	–	–	–
LVF	6.1	–	–	–
SD	3.0	3.0	4*	8.4
FA	11.5	7.0	–	c.18
AA	19.0	20.0	–	c.19
RQ	3.5	–	–	2.7*
SQ	6.2	–	–	7.0*
BI	4.1	4.5	–	4.4
L/MLD	5.2	–	–	6.5
FQ	1.2	1.1	–	1.0
MLD/LD	1.1	1.1	–	1.0

Table 62. Univariate analysis of *Belemnitella p. propinqua* from the Bavnodde Greensand. Measurements in millimetres.

Character	N	$\bar{X}$	SD	CV	OR
L	7	77.1	6.0	7.8	70.0–84.5
D	6	23.9	3.5	14.7	18.5–27.0
LAP	15	54.7	8.3	15.1	43.5–73.0
DVDP	17	12.5	1.6	12.6	9.7–16.0
LDP	17	12.9	1.7	13.2	9.8–16.4
MLD	14	14.2	1.8	12.3	11.3–17.2
MDVD	14	12.7	1.3	10.4	10.4–15.2
DVDAE	5	13.4	1.2	9.2	12.2–14.8
LDAE	5	12.9	1.1	8.2	11.4–14.3
SD	13	4.7	1.2	4.6	3.1–7.1
FA	9	16.1	8.5	53.1	9.0–38.0
RQ	6	3.3	0.6	17.5	2.7–4.3
SQ	5	5.8	0.3	4.9	5.4–6.2
BI	15	not calculated			3.5–5.5
L/MLD	7	not calculated			4.9–6.7
FQ	14	1.1	0.04	3.9	1.1–1.2
MLD/LDP	14	1.1	0.09	8.1	1.0–1.3

Table 63. Estimates of statistical parameters of two regression analyses of *Belemnitella p. propinqua* from the Bavnodde Greensand. Measurements in millimetres.

$y=a+bx$	N	r	Probability	SD_a	SD_b	SD_{yx}	t_a	Probability
DVDP=7.3675+0.0973LAP	15	0.5461	$0.05>P>0.02$	2.3230	0.0420	1.2980	3.1716	$0.01>P>0.005$
LDP=2.5921+0.7407MLD	14	0.8664	$P<0.001$	1.8629	0.1303	0.8218	1.3914	$0.20>P>0.10$

of the variates is 1.1. Owing to individual variation, this ratio varies from 1.0 to 1.3.

The holotype of *B. p. propinqua*, another specimen from the Eriksdal marl, and GSC 99475 from Russia are also plotted in Figs. 70–71.

Discussion. – Christensen (1974, unpublished) made univariate and bivariate biometric analyses of a sample of *B. p. propinqua* from Särdal, southern Sweden. This sample is Lower Santonian. The results are reported in Table 64 for comparison.

The correlation coefficient is very highly significant, $P<0.001$, with 15 degrees of freedom. The t-test on the y-intercept gives a value of 0.0996 with 15 degrees of freedom, which is not significant ($P>0.90$), implying an isometric relationship of the variates. The scatter plot of the length from the apex to the protoconch *vs.* the dorsoventral diameter at the protoconch is also shown in Fig. 70, as is the regression line.

The relationship of the length from the apex to the protoconch *vs.* the dorsoventral diameter at the protoconch thus differs in the two samples. It is isometric in the sample from Särdal but strongly allometric in the sample from the Bavnodde Greensand. This may be explained in the following way. The sample from Särdal mainly consists of juvenile specimens and the growth curve of these does not differ significantly from isometry. The sample from the Bavnodde Greensand mainly consists of adult specimens, and the growth curve of these is allometric.

An inspection of the scatter plot of *B. p. propinqua* from Särdal and the Bavnodde Greensand shows that the trend of the points is curved (Fig. 70). The power function, $y=bx^a$, also referred to as the equation of the simple allometry, was calculated on the basis of the specimens of the two samples. DVDP=0.41LAP$^{0.86}$; $N=32$; coefficient of determination=80%.

The power curve is also shown in Fig. 70, and it is evident that the power curve approximates the data well. The growth of *B. p. propinqua* is thus allometric. It is also clear that the growth curve calculated on the basis of the juvenile specimens from Särdal is very close to the lower half of the power curve.

Because of the allometric growth, the Birkelund Index changes with the size. If the coefficient of determination is 1.000, the Birkelund Index will vary as shown in Table 65. Adult specimens are thus more slender than juvenile

specimens. Owing to individual variation, the Birkelund Index varies from 3.3 (a juvenile specimen) to 5.5 (an adult specimen).

The sample from Särdal is very closely similar to the sample from the Bavnodde Greensand with respect to the internal characters and depth of pseudoalveolus.

B. p. propinqua shows affinity to the genus *Goniocamax* as used herein. Some specimens possess goniocamacoid characters, including the presence of a pseudoalveolus and the markedly ventral flattening of the guard, which is lanceolate in ventral view, e.g., the holotype from the upper Middle Santonian Eriksdal marl (Pl. 1:15) and specimens from the Bavnodde Greensand (Pl. 3:11; Fig. 69B). Other specimens display belemnitelloid characters and have a true alveolus, are subcylindrical in ventral view and conical in lateral view, and are only slightly flattened ventrally, e.g., the specimen from the Eriksdal marl figured as *A. propinquus?* by Moberg (1885, Pl. 6:22) (Pl. 1:16 herein) and specimens from the Bavnodde Greensand (Figs. 69A, C). Since these forms occur together in the Lower and Middle Santonian and are connected by all intermediate forms, they are considered as morphological varieties of the same species.

B. p. propinqua is most likely the direct successor of *B. schmidi* sp.nov. The affinity to this species is discussed above. *B. p. propinqua* is larger and stouter, and has a deeper pseudoalveolus and more distinct vascular markings than *G. mirabilis*.

The concept of *B. p. propinqua* was misinterpreted by Russian palaeontologists, including Arkhangelsky (1912), Jeletzky (1949), Nikitin (1958), Naidin (1964a, 1964b, 1974), and Glazunova (1972) (see Christensen 1971, 1991). The reason for that may be that Moberg (1885) erroneously noted that the depth of the pseudolveolus of the holotype is 17 mm. Christensen (1971, p. 374) showed, however, that this is incorrect, because the bottom of the pseudoalveolus is infilled with sediment (Pl. 1:15D). The depth of the pseudoalveolus is in fact 22.3 mm. Another reason may be that the holotype displays goniocamacoid characters.

Naidin (1974) employed a different concept of *B. p. propinqua* than that of Christensen (1971, 1973, 1986, 1991, and herein). He recognized *B. p. propinqua* from the Lower Santonian and *B. propinqua rylskiana* from the upper Lower to Upper Santonian, in addition to the dubious *B. propinqua mirabilis* Arkhangelsky, 1912 (=*Gonio-*

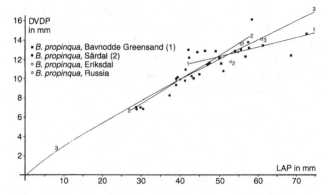

Fig. 70. Scatter plots and regression lines of the length from the apex to the protoconch (LAP) *vs.* the dorsoventral diameter at the protoconch (DVDP) of *Belemnitella propinqua propinqua* from the Bavnodde Greensand and Särdal. Regression line 1 is calculated on the basis of material from the Bavnodde Greensand and line 2 on the basis of material from Särdal. Growth curve 3 is the power curve, which is calculated on the basis of specimens from the Bavnodde Greensand and Särdal. 1, SGU Type 3906, holotype (Pl. 1:15, herein); 2, specimen figured as *Actinocamax propinquus?* by Moberg (1885, Pl. 6:22) (Pl. 1:16, herein); and 3, GSC 99475 (Fig. 69D, herein). + = mean values.

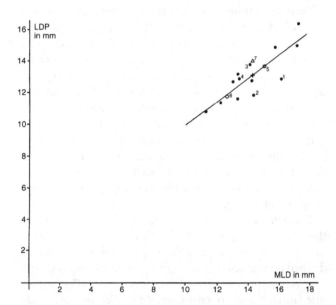

Fig. 71. Scatter plot and regression line of the maximum lateral diameter (MLD) *vs.* the lateral diameter at the protoconch (LDP) of *Belemnitella propinqua propinqua* from the Bavnodde Greensand (dots). Two specimens from the Eriksdal marl (open circles) and one specimens from Russia (open triangle) are also plotted. 1, GPIK 3921 (Pl. 3:11); 2, MGUH 23737 (Fig. 69B); 3, MGUH 23738 (Fig. 69C); 4, MGUH 23736 (Fig. 69A); 5, holotype, (Pl. 1:15); 6, specimen figured as *Actinocamax propinquus?* by Moberg (1885, Pl. 6:22) (Pl. 1:16 herein); 7, GSC 99475 (Fig. 69D). + = mean values.

camax mirabilis) from northern Kazakhstan. The latter taxon is discussed above. According to Naidin, *B. propinqua rylskiana* is distinguished from *B. p. propinqua* by its shape of the guard, depth of pseudoalveolus, fissure angle, and Schatzky distance (Table 66). This classification is not

Table 64. Univariate and bivariate analyses of *Belemnitella p. propinqua* from Särdal, southern Sweden.

Univariate analysis

Character	N	X̄	SD	CV	OR
L	4	57.2	11.8	20.7	40.3–65.6
D	4	17.7	5.8	32.6	10.3–23.6
LAP	17	40.1	7.5	18.8	28.9–57.6
SD	24	5.4	0.8	15.4	3.7–6.7
FA	27	16.8	6.5	39.0	10.0–33.7
RQ	4	3.4	0.5	14.7	2.8–3.9
SQ	4	6.2	0.2	3.2	6.0–6.4
BI	17	4.1	0.3	8.2	3.3–4.5

Bivariate analysis

DVDP = −0.1283 + 0.2485 LAP; $N=17$; $r=0.8971$;
$SD_a = 1.2876$; $SD_b = 0.0316$; $SD_{yx} = 0.9525$

Table 65. Tabulation of the Birkelund Index (BI) with increasing length from the apex to the protoconch (LAP) in *Belemnitella p. propinqua*. Owing to allometry, the Birkelund Index will increase when the length increases. See text for discussion.

LAP (mm)	BI
30	3.9
40	4.1
50	4.2
60	4.3
70	4.4
75	4.5

Table 66. Tabulation of differences between *Belemnitella p. propinqua* (Moberg) and *B. propinqua rylskiana* (Nikitin) according to Naidin (1974). This subspecies concept is not considered tenable (see text for full discussion).

Naidin (1974)	*B. p. propinqua*	*B. p. rylskiana*
Age	Lower Santonian	upper Lower–Upper Santonian
Shape in ventral view	lanceolate	cylindrical
Shape in lateral view	slightly lanceolate or cigar shaped	conical
Fissure angle	'large'	'small'
Schatzky distance	2.5–4.5 mm	5.5–7.0 mm
Riedel-Quotient	4–5	3

tenable, because *B. p. propinqua sensu* Naidin and *B. propinqua rylskiana sensu* Naidin occur together in the Lower Santonian Bavnodde Greensand and upper Middle Santonian Eriksdal marl (see above). *B. propinqua rylskiana* is therefore placed in synonymy with *B. p. propinqua*. Moreover, *B. p. propinqua* should for obvious reasons be interpreted with respect to its holotype, which is upper Middle Santonian. The Riedel-Quotient of the holotype is 3.5, which is very close to the mean values of the Riedel-Quotient of the two samples from the Bavnodde Greensand and Särdal. The Riedel-Quotient of the holotype is outside the range of *B. p. propinqua sensu* Naidin. The taxonomical position of *B. p. propinqua sensu* Naidin (1974) is uncertain. It may include *B. p. propinqua* in the sense used here as well as *B. schmidi* sp. nov.

Distribution. – *B. p. propinqua* occurs in southern Sweden, on Bornholm, and on the Russian Platform. Outside this area it has been recorded only from southern England (Christensen 1991). It appears some way above the base of the Santonian and continues into the upper Middle Santonian. Jeletzky (1955) noted that GSC 99475 came from the *Inoceramus* ex gr. *I. cardissoides* Zone *sensu rossico*; that is probably Lower Santonian.

Summary

Belemnites from the Lower and Middle Coniacian Arnager Limestone Formation and the Upper Coniacian and Lower Santonian Bavnodde Greensand Formation are described using univariate and bivariate biometrical methods. The study is based mainly on material collected recently. In addition, collections made in the last part of the 19th century and the beginning of this century are also included. The formations and outcrops are placed in the international stratigraphical framework on the basis of ammonites, inoceramid bivalves, foraminifera, dinoflagellate cysts as well as belemnites. The Arnager Limestone at its type locality has yielded a diverse fauna of lower Lower Coniacian inoceramid bivalves. It is therefore enigmatic that the Middle Coniacian index ammonite, *Peroniceras tridorsatum*, also is recorded from the type locality. The marl of the Arnager Limestone Formation at Muleby Å is upper Lower Coniacian, and the sandy marl of the Arnager Limestone Formation at Stampe Å is lower Middle Coniacian on inoceramid bivalve evidence. The Bavnodde Greensand Formation is Upper Coniacian and Lower Santonian on ammonite, belemnite, and inoceramid bivalve evidence. The Upper Cretaceous sedimentary succession of Bornholm is about 200 m thick. The lower Middle Cenomanian Arnager Greensand Formation is about 85 m thick, and the Bavnodde Greensand Formation is about 70 m thick. The Arnager Limestone Formation, which is sandwiched between the two greensands, is 12–20 m thick at the type locality, and the thickness increases to about 35–40 m in the Stampe Å area.

The belemnite fauna of the Arnager Limestone Formation includes four species: *Actinocamax verus* cf. *antefragilis*, new for Bornholm, *Goniocamax l. lundgreni*, *Goniocamax* sp. nov. aff. *lundgreni*, and *Goniocamax*(?) sp. nov. (?). The fauna of the Bavnodde Greensand Formation is remarkably diverse and comprises nine species: *Actinocamax v. verus*, *Gonioteuthis praewestfalica*, new for Bornholm, *Gonioteuthis w. westfalica*, *Gonioteuthis ernsti* sp. nov., *Goniocamax l. lundgreni*, *Goniocamax birkelundae* sp. nov., *Goniocamax striatus* sp. nov., *Belemnitella schmidi* sp. nov., and *Belemnitella p. propinqua*. Two subgenera of *Actinocamax* are recognized: *A. (Actinocamax)* and *A. (Praeactinocamax)*. The first subgenus is characterized by the small size of the guard, usually long cone-shaped alveolar fracture, and isometric growth. The latter subgenus is distinguished by the medium to large size of the guard, usually short cone-shaped alveolar fracture, and allometric growth. *Goniocamax*, type species *Actinocamax lundgreni*, was established as a subgenus of *Gonioteuthis* and is here raised to the rank of a genus. It includes the type species and its close relatives, *G. birkelundae* sp. nov., *G. striatus* sp. nov., *G. esseniensis* (Christensen), and *G. mirabilis* Arkhangelsky.

Three belemnite zones and eight assemblage zones are established. The bases of most of the zones cannot be precisely defined, owing to minor stratigraphical gaps between the exposures. The Coniacian has a Zone of *G. lundgreni*, which is subdivided into three assemblage zones. The Lower Santonian has a Zone of *G. praewestfalica* below and a Zone of *G. westfalica* above. The lower Zone includes two assemblage zones and the upper zone comprises three assemblage zones. The zones are correlated with the faunal zones and *Gonioteuthis* Zones of Lägerdorf, NW Germany, as well as the inoceramid bivalve assemblage zones of the North Temperate palaeobiogeographical Province.

The belemnite faunas of Bornholm include species of the Central European and Central Russian Subprovinces of the North Temperate palaebiogeographic Province as defined on the basis of belemnites. There is a major faunal turnover at the base of the *G. westfalica* Zone. The belemnite faunas below are characterized by *Goniocamax lundgreni* with subordinate species of *Actinocamax*, the *Gonioteuthis* lineages, and the *Goniocamax–Belemnitella* lineage. The faunas above are characterized by *Gonioteuthis westfalica* with subordinate species of *Actinocamax* and the *Goniocamax–Belemnitella* lineage. Bornholm was part of the Central Russian Subprovince in the Coniacian and lower Lower Santonian and became part of the Central European Subprovince in the remaining part of the Santonian. The evolutionary trends of Coniacian and Santonian species of the genera *Goniocamax*, *Gonioteuthis*, and *Belemnitella* from NW Europe are discussed, with special reference to Bornholm.

The stratigraphical distribution of the genus *Actinocamax* is punctuated. *A. verus* cf. *antefragilis* occurs in the upper Lower and Middle Coniacian of Bornholm, and *A.*

v. verus enters in the upper Lower Santonian. The genus is not recorded from the upper part of the Coniacian and the lower and middle Lower Santonian. *A. verus* cf. *antefragilis* was recorded earlier only from the Lower Turonian of the Russian Platform.

Gonioteuthis praewestfalica occurs in the upper Middle and Upper Coniacian and *G. westfalica* in the upper Lower and Lower Middle Santonian of the white chalk of Lägerdorf, NW Germany. The upper boundary of *G. praewestfalica* has not been established at Lägerdorf, because *Gonioteuthis* is extremely rare in the lower Lower Santonian *pachti/undulatoplicatus* Zone. On Bornholm, *G. praewestfalica* is distributed in the lower Lower Santonian and is followed upwards by *G. westfalica* in the middle and upper Lower Santonian. The upper range of *G. praewestfalica* is thus extended into the lower Lower Santonian. Samples of *G. praewestfalica* and *G. westfalica* from the off-shore white chalk characteristically include only adult specimens, whereas samples from the nearshore greensands of Bornholm comprise all growth stages, indicating that the species reproduced in shallow waters close to the Fennoscandian Shield.

Gonioteuthis ernsti sp.nov. enters in the upper Lower Santonian and is recorded from Bornholm and the Münster Basin. It probably evolved from *G. westfalica* by allopatric speciation.

Goniocamax lundgreni, which appears at the base of the Lower Coniacian, is considered as the earliest member of *Goniocamax*. It appears later on the Russian Platform than on Bornholm, and we therefore suggest that this taxon migrated to the Russian Platform from Baltoscandia. Four subspecies of *G. lundgreni* are discussed in detail and placed in synonymy with *G. l. lundgreni*, as are *Actinocamax* aff. *westfalicus sensu* Birkelund (1957) and *A. propinquus* mut. (var.) nov. Stolley (1897).

Goniocamax birkelundae sp.nov. enters near the base of the Upper Coniacian. It probably evolved by allopatric speciation from *G. lundgreni*. *Goniocamax striatus* sp.nov. enters simultaneously with *Actinocamax v. verus* near the base of the upper Lower Santonian. It is most likely the lineal descendant of *G. birkelundae*.

We consider *Belemnitella schmidi* sp.nov., which appears at the base of the Lower Santonian, as the earliest member of *Belemnitella*. It is a transition form possessing characters in common with the genera *Goniocamax* and *Belemnitella*. It probably evolved by allopatric speciation from *Goniocamax lundgreni*. *Belemnitella p. propinqua* enters some way above the base of the Lower Santonian and is probably the lineal descendant of *B. schmidi* sp.nov. *B. p. propinqua* is interpreted on the basis of its holotype, and this concept differs from that used by Russian palaeontologists.

Acknowledgements. – The majority of the belemnites examined during the course of the present study were collected by the authors during joint field work in 1975–1977, 1984, and 1993. WKC did additional field work in 1979, 1983, and 1986–1987. Professor F. Schmid, Worblingen, Dr. H. Ernst, Hamburg, and Mr. S.L. Jakobsen, Geological Museum, Copenhagen, participated in the field work in various years. We are most grateful for this help. Additional belemnites were kindly provided by colleagues from the Geological Institute, University of Copenhagen, Geological Institute, University of Århus, Geological Survey of Denmark, Copenhagen, and Geologisch-Paläontologisches Institut der Universität Hamburg.

We thank the curators and staff of the following institutes and museums, who allowed us to study belemnites in their care: Geological Survey of Sweden, Uppsala; Geological Institute, University of Lund, Sweden; Swedish Museum of Natural History, Section of Paleozoology, Stockholm, Sweden; Geologisch-Paläontologisches Institut, Greifswald, Germany; and Geological Survey of Canada, Ottawa. J. Mutterlose (Bochum) reviewed the manuscript. We also thank Mr. S.L. Jakobsen, who made the photographs with great skill and care, and the Geological Institute, Copenhagen for loan of the field station Gravgærde on the island of Bornholm. The field work of WKC was supported by the Geological Museum and the Carlsberg Foundation.

References

Arkhangelsky, A.D. 1912: Verkhnemelovye otlozhenija vostoka Evropejskoj Rossii. [The Upper Cretaceous deposits in the eastern part of the Europen Russia.] *Materialy dlya Geologii Rossii 25.* 631 pp. [In Russian.]

Bailey, H.W. & Hart, M.B. 1979: The correlation of the Early Senonian in western Europe using Foraminifera. *In* Wiedmann, J. (ed.): *Aspekte der Kreide Europas. International Union of Geological Sciences A6,* 159–169.

Bailey, H.W., Gale, A.S., Mortimore, R.N., Swiecicki, A. & Wood, C.J. 1983: The Coniacian–Maastrichtian Stages of the United Kingdom, with particular reference to southern England. *Newsletters on Stratigraphy 12,* 29–42.

Bailey, H.W., Gale, A.S., Mortimore, R.N., Swiecicki, A. & Wood, C.J. 1984: Biostratigraphical criteria for the recognition of the Coniacian to Maastrichtian stage boundaries in the Chalk of north-west Europe, with particular reference to southern England. *Bulletin of the Geological Society of Denmark 33,* 31–39.

Bandel, K. & Spaeth, C. 1988: Structural differences in the ontogeny of some belemnite rostra. *In* Wiedmann, J. & Kullmann, J. (eds): *Cephalopods – Present and Past,* 247–271. Schweizerbart'sche, Stuttgart.

Bayle, E. 1878: Fossiles principaux des terrains de la France. *Explication de la Carte Géologique de la France 4:1,* Atlas. 79 pls.

Birkelund, T., 1957: Upper Cretaceous belemnites of Denmark. *Biologiske Skrifter. Det Kongelige Danske Videnskabernes Selskab 9.* 69 pp.

Blainville, H.M.D. de 1825–1827: *Manual de Malacologie et de Conchyliologie.* 664 pp. (1825), 87 plates (1827). Levrault, Paris.

Blainville, H.M.D. de 1827: *Mémoire sur les Belemnites, considerées zoologiquement et géologiquement.* 136 pp. Levrault, Paris.

[Brüsch, W. 1984: Bavnodde Grønsandets palæomiljø og -geografi. Unpublished M.Sc. Thesis. 181 pp. Copenhagen University.]

Christensen, W.K. 1971: *Belemnitella propinqua propinqua* (Moberg, 1885) from Scandinavia. *Bulletin of the Geological Society of Denmark 20,* 369–384.

Christensen, W.K. 1973: The belemnites and their stratigraphical significance. *In* Bergström, J., Christensen, W.K., Johansson, C. & Norling, E.: An extension of Upper Cretaceous rocks to the Swedish west coast at Särdal. *Bulletin of the Geological Society of Denmark 22,* 113–140.

Christensen, W.K. 1974: Morphometric analysis of *Actinocamax plenus* from England. *Bulletin of the Geological Society of Denmark 23,* 1–26.

Christensen, W.K. 1975a: Upper Cretaceous belemnites from the Kristianstad area in Scania. *Fossils and Strata 7.* 69 pp.

Christensen, W.K. 1975b: Designation of lectotypes for *Gonioteuthis westfalicagranulata* and *G. granulataquadrata. Paläontologische Zeitschrift 50,* 126–134.

Christensen, W.K. 1976: Palaeobiogeography of Late Cretaceous belemnites of Europe. *Paläontologische Zeitschrift 50*, 113–129.

Christensen, W.K. 1982: Late Turonian – early Coniacian belemnites from western and central Europe. *Bulletin of the Geological Society of Denmark 31*, 63–79.

Christensen, W.K. 1985 [date of imprint 1984]: The Albian to Maastrichtian of southern Sweden and Bornholm, Denmark: a review. *Cretaceous Research 5*, 313–327.

Christensen, W.K. 1986. Upper Cretaceous belemnites from the Vomb Trough in Scania, Sweden. *Sveriges Geologiska Undersökning Ca 57.* 57 pp.

Christensen, W.K. 1988: Upper Cretaceous belemnites of Europe: State of the art. *In* Streel, M. & Bless, M.J.M. (eds): *The Chalk District of the Euregio Meuse–Rhine*, 5–16. Natuurhistorisch Museum, Maastricht, and Laboratoires de Paléontologie de l'Université d'Etat, Liège.

Christensen, W.K. 1990a: *Actinocamax primus* Arkhangelsky (Belemnitellidae; Upper Cretaceous): Biometry, comparison and biostratigraphy. *Paläontologische Zeitschrift 64*, 75–90.

Christensen, W.K. 1990b: Upper Cretaceous belemnite stratigraphy of Europe. *Cretaceous Research 11*, 371–386.

Christensen, W.K. 1991: Belemnites from the Coniacian to Lower Campanian chalks of Norfolk and southern England. *Palaeontology 34*, 695–749.

Christensen, W.K. 1993a: Upper Cretaceous belemnitellids from the Båstad Basin, southern Sweden. *Geologiska Föreningens i Stockholm Förhandlingar 115*, 39–57.

Christensen, W.K. 1993b: *Actinocamax cobbani* n.sp. from the Coniacian of Montana and Wyoming and the occurrence of Late Cretaceous belemnites in North America and Greenland. *Journal of Paleontology 67*, 434–446.

Christensen, W.K. 1994: Upper Cretaceous belemnites from Lonzée (SE Belgium) and their stratigraphic significance. *Bulletin de l'Institut Royal des Sciences Naturelles de Belgique, Sciences de la Terre 64*, 151–158.

Christensen, W.K. 1995: *Belemnitella* from the Upper Campanian and Lower Maastrichtian chalk of Norfolk, England. *Special Papers in Palaeontology 51.* 84 pp.

Christensen, W.K. & Hoch, E. 1983: *Actinocamax* cf. *manitobensis* from the Kangerdlugssuaq area, southern East Greenland. *Bulletin of the Geological Society of Denmark 32*, 33–42.

Christensen, W.K. & Schmid, F. 1987: The belemnites of the Vaals Formation from the C.P.L. Quarry at Hallembaye in Belgium – Taxonomy, biometry and biostratigraphy. *Geologisches Jahrbuch A94*, 3–37.

Cieslinski, S. 1963: Die Grundlagen der Biostratigraphie der Oberkreide in Polen. *Bericht der Geologischen Gesellschaft in der Deutschen Demokratischen Republik 8*, 189–197.

Douglas, R.G. & Rankin, C. 1969: Cretaceous planktonic foraminifera from Bornholm. *Lethaia 2*, 185–219.

Ernst, G. 1963a: Stratigraphische und gesteinschemische Untersuchungen im Santon und Campan von Lägerdorf (SW-Holstein). *Mitteilungen aus dem Geologischen Staatsinstitut in Hamburg 32*, 71–127.

Ernst, G. 1963b: Zur Feinstratigraphie und Biostratonomie des Obersanton und Campan von Misburg und Höver bei Hannover. *Mitteilungen aus dem Geologischen Staatsinstitut in Hamburg 32*, 128–147.

Ernst, G. 1964: Ontogenie, Phylogenie und Stratigraphie der Belemnitengattung *Gonioteuthis* Bayle aus dem nordwestdeutschen Santon/ Campan. *Fortschritte in der Geologie von Rheinland und Westfalen 7*, 113–174.

Ernst, G. 1966: Fauna, Ökologie und Stratigraphie der mittelsantonen Schreibkreide von Lägerdorf (SW Holstein). *Mitteilungen aus dem Geologischen Staatsinstitut in Hamburg 35*, 115–150.

Ernst, G. 1968: Die Oberkreide-Aufschlüsse im Raume Braunschweig-Hannover und ihre stratigraphische Gliederung mit Echinodermen und Belemniten. 1. Teil: Die jüngere Oberkreide (Santon–Maastricht). *Beihefte zu den Berichten der Naturhistorischen Gesellschaft zu Hannover 5*, 235–284.

Ernst, G. & Schulz, M.-G. 1974: Stratigraphie und Fauna des Coniac und Santon im Schreibkreide-Richtprofil von Lägerdorf (Holstein). *Mitteilungen aus dem Geologisch-Paläontologischen Institut der Universität Hamburg 43*, 5–60.

Glazunova, A.E. 1972: Paleontologicheskoye obosnovaniye stratigraficheskogo raschleneniya melovykh otlozhenij Povolzh'ya; verkhnij mel. [The paleontological basis for stratigraphical subdivision of the Cretaceous deposits of the Volga region; Upper Cretaceous]. *Ministry of the Geology of the USSR. Geological Institute VSEGEI.* 203 pp. [In Russian.]

Gravesen, P., Rolle, F. & Surlyk, F. 1982: Lithostratigraphy and sedimentary evolution of the Triassic, Jurassic and Lower Cretaceous of Bornholm. *Danmarks Geologiske Undersøgelse B7.* 51 pp.

Gry, H. 1960: Geology of Bornholm – Guide to Excursions Nos A 45 and C 40. *International Geolological Congress XXI Session, Norden.* Copenhagen, 16 pp.

Gry, H. 1969: Megaspores from the Jurassic of the island of Bornholm, Denmark. *Meddelelser fra Dansk Geologisk Forening 19*, 69–89.

Hald, A. 1957: *Statistical Theory with Engineering Applications.* 783 pp. Wiley, New York, N.Y.

Hamann, N.E. 1987: Bornholm. Geological map of Mesozoic formations. *The National Forest and Nature Agency.*

Hamann, N.E. 1988: Bornholms Geologi 1. Mesozoikum. *Varv 1988, 2*, 64–75.

Hamann, N.E. 1989: Bornholms Geologi IV. Mesozoikum. *Varv 1989, 3*, 74–104.

International Commission on Zoological Nomenclature 1985: Opinion 1328. *Belemnites mucronatus* Schlotheim, 1813 (Coleoidea): Conserved and neotype designated. *Bulletin of Zoological Nomenclature 42*, 222–225.

Janet, C. 1891: Note sur trois nouvelles bélemnites sénoniennes. *Bulletin de la Société Géologique de France 19*, 716–721.

Jarvis, I. 1980: Palaeobiology of Upper Cretaceous belemnites from phosphatic chalk of the Anglo-Paris Basin. *Palaeontology 23*, 889–914.

Jeletzky, J.A. 1948: Zur Kenntnis der Oberkreide der Dnjepr-Donez-Senke und zum Vergliech der russischen borealen Oberkreide mit derjenigen Polens und Nordwesteuropas. *Geologiska Föreningens i Stockholm Förhandlingar 70*, 583–602.

Jeletzky, J.A. 1949: Some notes on 'Actinocamax' propinquus Moberg 1885, its taxonomic position and phylogenetic relations within the family Belemnitellidae Pavlov 1913, morphological characters and synonymy. *Geologiska Föreningens i Stockholm Förhandlingar 71*, 415–424.

Jeletzky, J.A. 1950: *Actinocamax* from the Upper Cretaceous of Manitoba. *Geological Survey of Canada Bulletin 15.* 41 pp.

Jeletzky, J.A. 1955: Evolution of Santonian and Campanian *Belemnitella* and paleontological systematics: exemplified by *Belemnitella praecursor* Stolley. *Journal of Paleontology 29*, 478–509.

Jeletzky, J.A. 1958: Die jüngere Oberkreide (Oberconiac bis Maastricht) Südwestrusslands, und ihr Vergleich mit der Nordwest- und Westeuropas. *Beihefte zum Geologischen Jahrbuch 33.* 157 pp.

Jeletzky, J.A. 1961: *Actinocamax* from the Upper Cretaceous Benton and Niobrara Formations of Kansas. *Journal of Paleontology 35*, 505–531.

Jeletzky, J.A. 1965: Taxonomy and phylogeny of fossil Coleoidea (=Dibranchiata). *Geological Survey of Canada, Paper 65-2, No. 42*, 72–76.

Kaplan, U. & Kennedy, W.J. 1994: Ammoniten des westfälischen Coniac. *Geologie und Paläontologie in Westfalen 31.* 155 pp.

Kaplan, U., Kennedy, W.J. & Wright, C.W. 1987: Turonian and Coniacian Scaphitidae from England and North-West Germany. *Geologisches Jahrbuch A103*, 5–39.

Kauffman, E.G. 1973. Cretaceous Bivalvia. *In* Hallam, A. (ed.): *Atlas of Palaeobiography*, 353–383. Elsevier Scientific Publishing Company. Amsterdam.

Kennedy, W.J. 1984: Systematic palaeontology and stratigraphic distribution of the ammonite faunas of the French Coniacian. *Special Papers in Palaeontology 31.* 160 pp.

Kennedy, W.J. & Christensen, W.K. 1991: Coniacian and Santonian ammonites from Bornholm, Denmark. *Bulletin of the Geological Society of Denmark 38*, 203–226.

Kennedy, W.J. & Cobban, W.A. 1991: Coniacian ammonite faunas from the United States Western Interior. *Special Papers in Palaeontology 45*. 96 pp.

Kennedy, W.J., Hancock, J.M. & Christensen, W.K. 1981: Albian and Cenomanian ammonites from the island of Bornholm (Denmark). *Bulletin of the Geological Society of Denmark 29*, 203–244.

Kongiel, R. 1962: On belemnites from the Maastrichtian, Campanian and Santonian sediments in the Middle Vistula Valley (Central Poland). *Prace Muzeum Ziemi 5*. 148 pp.

Miller, J.S. 1823: Observations on the genus *Actinocamax*. *Transactions of the Geological Society (2), 2*, 45–62.

Moberg, J.C. 1885: Cephalopoderna i Sveriges Kritsystem. 2. Artsbeskrifning. *Sveriges Geologiska Undersökning C73*. 45 pp.

Naidin, D.P. 1952: Verkhnemelovye belemnity zapadnoj Ukrainy. [The Upper Cretaceous belemnites of western Ukraine.] *Trudy Moskovskogo Geologo-Razvedochnogo Instituta imeni S. Ordzhinikidze 27*. 170 pp. [In Russian.]

Naidin, D.P. 1953: Novyj belemnit iz verkhnemelovykh otlozhenij. [A new belemnite from the Upper Cretaceous deposits of Crimea.] *Byulleten' Moskovskogo Obshchestva Ispytatelej Prirody, Otdel Geologicheskij 28*, 64–65. [In Russian.]

Naidin, D.P. 1959. Subclass Endocochlia. Order Decapoda. Suborder Belemnoidea. *In* Moskvin, M.M (ed.): *Atlas Verkhnemelovoi Fauny Severnogo Kavkaza i Kryma [Atlas of the Upper Cretaceous Fauna of the northern Caucasus and Crimea]*, 198–209. Gostoptekhizdat, Moscow. [In Russian.]

Naidin, D.P. 1964a: Verkhnemelovye belemnity Russkoj platformy i nekotorych sopredel'nych oblastej. [Upper Cretaceous *Belemnitella* and *Belemnella* from the Russian Platform and some adjacent regions.] *Byulleten' Moskovskogo Obshchestva Ispytatelej Prirody, Otdel Geologicheskii 39*, 85–97. [In Russian.]

Naidin, D.P. 1964b: *Verkhnemelovye belemnity Russkoj platformy i sopredel'nykh oblastej; aktinokamaksy, gonioteitisy i belemnellokamaksy. [Upper Cretaceous belemnites of the Russian Platform and contiguous regions, Actinocamax, Gonioteuthis, Belemnellocamax.]* 190 pp. Moscow University Press, Moscow. [In Russian.]

Naidin, D.P. 1974: Subclass Endocochlia. Order Belemnitidae. *In* Krymgol'ts, G. Ya. (ed.): *Atlas verkhnemelovoj fauny Donbass [Atlas of the Upper Cretaceous fauna of Donbass]*, 197–240. 'Nedra', Moscow. [In Russian.]

Naidin, D.P. 1979: Belemnitellidy [Belemnitellids]. *In* Papulov, G.N. & Naidin, D.P. (eds): Granitsa santona i kampana na Vostochno-Evropejskoj platforme. [The Santonian–Campanian boundary in the East-European Platform.] *Trudy Instituta Geologii i Geokhimii AN SSSR 148*, 75–89. [In Russian.]

Naidin, D.P. 1981: The Russian Platform and Crimea. *In* Reyment, R.A. & Bengtson, P. (eds): *Aspect of Mid-Cretaceous Regional Geology*, 29–68. Academic Press, London.

Naidin, D.P. & Kopaevich, L.F. 1977: O zonal'nom delenii verkhnego mela Evropejskoj paleobiogeograficheskoj oblasti. [The zonation of the Upper Cretaceous of the European palaeobiogeographical province.] *Byulleten' Moskovskogo Obshchestva Ispytatelej Prirody, Otdel Geologicheskij 52*, 92–112. [In Russian.]

Nikitin, I.I. 1958: Verkhnemelovye belemnity severo-zapadnogo kryla Dneprovsko-Donetskoj vpadiny. [Upper Cretaceous belemnites of the northeastern slope of the Dnjepr–Donez-Basin.] *Trudy Instituta Geologicheskikh Akademia Nauk Ukrainskoi SSR, Seriya Stratigrafii i Paleontologii 20*. 92 pp. [In Ukrainian.]

Nilsson, S. 1826: Om de mångrummiga snäckor som förekomma i kritformationen i Sverige. *Kungliga Svenska Vetenskapsakademiens Handlingar 1825*, 329–343.

Noe-Nygaard, N. & Surlyk, F. 1985: Mound bedding in a sponge-rich Coniacian chalk, Bornholm, Denmark. *Bulletin of the Geological Society of Denmark 34*, 237–249.

Orbigny, A. d' 1840: *Paléontologie francaise. Terrains Crétacés.* 642 pp. Paris.

Packer, S.R. & Hart, M.B. 1994: Evidence for sea level change from the Cretaceous of Bornholm, Denmark. *GFF 116*, 167–173.

Packer, S., Hart, M.B., Tocher, B.A. & Braley, S. 1989: Upper Cretaceous microbiostratigraphy of Bornholm, Denmark. *In* Batten, D.J. & Keen, M.C. (eds): *Northwest European Micropalaeontology and Palynology*, 236–247. Ellis Horwood, Chicester.

Pavlow, A.P. 1914: Jurskie i Nizhnemelovye Cephalopopda severnoi Sibiri. [Jurassic and Lower Cretaceous Cephalopoda of northern Siberia.] *Imperatovski Akademia Nauk, St Petersburg, Zapiski Seriia 8, Fiziko- mathematicheskii 21*. 68 pp. [In Russian.]

Ravn, J.P.J. 1916: Kridtaflejringerne paa Bornholms Sydvestkyst og deres fauna. I. Cenomanet. *Danmarks Geologiske Undersøgelse, II Række, 30.* 40 pp.

Ravn, J.P.J. 1918: Kridtaflejringerne paa Bornholms sydvestkyst og deres fauna. II. Turonet. *Danmarks Geologiske Undersøgelse, II Række, 31.* 39 pp.

Ravn, J.P.J. 1921: Kridtaflejringerne paa Bornholms sydvestkyst og deres fauna. III. Senonet. IV. Kridtaflejringerne ved Stampe Aa. *Danmarks Geologiske Undersøgelse, II Række, 32.* 52 pp.

Ravn, J.P.J. 1930: Nogle bemærkninger om Bornholms kridtaflejringer. *Geologiske Föreningens i Stockholm Förhandlingar 52*, 279–283.

Ravn, J.P.J. 1946: Om Nyker omraadets kridtaflejringer. *Biologiske Skrifter. Det Kongelige Danske Videnskabernes Selskab 4.* 36 pp.

Reid, R.E.H. 1971: The Cretaceous rocks of north-eastern Ireland. *Irish Naturalists' Journal 17*, 105–129.

Reyment, R.A. & Naidin, D.P. 1962: Biometric study of *Actinocamax verus* s.l. from the Upper Cretaceous of the Russian Platform. *Stockholm Contributions in Geology 9*, 147–206.

Rohlf, F.J. & Sokal, R.R. 1969: *Statistical Tables.* 253 pp. Freeman, San Francisco, Cal.

Schiøler, P. 1992: Dinoflagellate cysts from the Arnager Limestone Formation (Coniacian, Late Cretaceous), Bornholm, Denmark. *Review of Palaeobotany and Palynology 72*, 1–25.

Schlotheim, E.F. von 1813: Beiträge zur Naturgeschichte der Versteinerungen in geognostischer Hinsicht. *Leonhard's Taschenbuch für die gesammte Mineralogi mit Hinsicht auf die neuesten Entdeckungen, Jahrgang 7.* 134 pp.

Schlüter, C. 1874: Die Belemniten der Insel Bornholm. *Zeitschrift der Deutschen geologischen Gesellschaft 26*, 827–855.

Schlüter, C. 1876: Cephalopoden der oberen deutschen Kreide. Teil 2. *Palaeontographica 24*, 123–263.

Schönfeld, J. 1990: Zur Stratigraphie und Ökologie benthischer Foraminiferen im Schreibkreide-Richtprofil von Lägerdorf/Holstein. *Geologisches Jahrbuch A117.* 151 pp.

Schulz, M.-G. 1979: Morphometrisch-variationsstatistische Untersuchungen zur Phylogenie der Belemniten-Gattung *Belemnella* im Untermaastricht NW-Europas. *Geologisches Jahrbuch A47.* 157 pp.

Schulz, M.-G. 1996: Macrofossil biostratigraphy. *In* Schönfeld, J. & Schulz, M.-G. (coord.) *et al.*: New results on biostratigraphy, palaeomagnetism, geochemistry and correlation from the standard section for the Upper Cretaceous White Chalk of northern Germany (Lägerdorf–Kronsmoor–Hemmoor). *Mitteilungen aus dem Geologisch-Paläontologischen Institut der Universität Hamburg 77*, 548–550.

Schulz, M.-G., Ernst, G., Ernst, H. & Schmid, F. 1984: Coniacian to Maastrichtian stage boundaries in the standard section for the Upper Cretaceous white chalk of NW Germany (Lägerdorf–Kronsmoor–Hemmoor): Definitions and proposals. *Bulletin of the Geological Society of Denmark 33*, 203–215.

Seitz, O. 1961: Die Inoceramen des Santon von Nordwestdeutschland. I. Teil (Die Untergattungen *Platyceramus*, *Cladoceramus* und *Cordiceramus*). *Beihefte zum Geologischen Jahrbuch 46.* 186 pp.

Seitz, O. 1965: Die Inoceramen des Santon und Unter-Campan von Nordwestdeutschland. II Teil, Biometrie, Dimorphismus und Stratigraphie der Untergattung *Sphenoceramus*. *Beihefte zum Geologischen Jahrbuch 69.* 194 pp.

Siegel, S. 1956: *Nonparametric Statistics for the Behavioral Sciences.* 312 pp. McGraw–Hill, New York, N.Y.

Sinzow, I. 1915: O verkhnemelovye osadkakh saratovskoj gubernii. [Upper Cretaceous sediments of the Saratow Province.] *Zapiski Imperatorskago Mineralogicheskogo Obshchestva (2), 50,* 133–162. [In Russian.]

Sokal, R.R. & Rohlf, F.J. 1969: *Biometry – The Principles and Practise of Statistics in Biological Research.* 776 pp. Freeman, San Francisco, Cal.

Solakius, N. 1988: The type material of the Upper Cretaceous benthic foraminifer *Pseudovalvulineria vombensis* Brotzen, 1945. *Geologiska Föreningens i Stockholm Förhandlingar 110,* 197–201.

Solakius, N. 1989: Foraminifera from the Arnager Limestone–Bavnodde Greensand boundary on Bornholm, Denmark. *Geologiska Föreningens i Stockholm Förhandlingar 111,* 101–104.

Solakius, N. & Larsson, K. 1985: Foraminifera and biostratigraphy of the Arnager Limestone, Bornholm, Denmark. *Danmarks Geologiske Undersøgelse C5.* 41 pp.

Stolley, E. 1896: Einige Bemerkungen über die obere Kreide, insbesondere von Lüneburg und Lägerdorf. *Archiv für Anthropologie und Geologie Schleswig–Holsteins 1,* 139–176.

Stolley, E. 1897: Ueber die Gliederung des norddeutschen und baltischen Senon sowie die dasselbe characterisierenden Belemniten. *Archiv für Antropologie und Geologie Schleswig–Holsteins 2,* 216–302.

Stolley, E. 1916: Neue Beiträge zur Kenntnis der norddeutschen oberen Kreide. III. Die Bedeutung der *Actinocamax*-Arten als Leitfossilien der oberen Kreide. *Jahresbericht des Niedersächsischen geologischen Vereins zu Hannover 9,* 95–104.

Stolley, E. 1930: Einige Bemerkungen über die Kreide Südskandinaviens. *Geologiske Föreningens i Stockholm Förhandlingar 52,* 157–190.

Tröger, K.-A. 1989: Problems of Upper Cretaceous inoceramid biostratigraphy and paleobiogeography in Europe and western Asia. *In* Wiedmann, J. (ed.): *Cretaceous of the Western Tethys,* 911–930. Schweizerbart'sche Verlagsbuchhandlung, Stuttgart.

Tröger, K.-A. & Christensen, W.K. 1991: Upper Cretaceous (Cenomanian–Santonian) inoceramid bivalve faunas from the island of Bornholm, Denmark. *Danmarks Geologiske Undersøgelse A28.* 47 pp.

Ulbrich, H. 1971: Mitteilungen zur Biostratigraphie des Santon und Campan des mittleren Teils der Subhercynen Kreidemulde. *Freiberger Forschungshefte C267,* 47–71.

Walaszczyk, I. 1992: Turonian through Santonian deposits of the Central Polish Uplands; their facies development, inoceramid paleontology and stratigraphy. *Acta Geologica Polonica 42.* 142 pp.

Wood, C.J. & Schmid, F. 1991: Upper Cretaceous of Helgoland (NW Germany): Lithology, palaeontology and biostratigraphy. *Geologiches Jahrbuch A120,* 37–61.

Zittel, K.A. von 1895: *Grundzüge der Palaeontologie (Paläozoologie).* 971 pp. Oldenburg, Munich.

Plates

Plate 1

All specimens are coated with ammonium chloride, and figures are all natural size unless otherwise stated.

□1. *Actinocamax verus verus* Miller, GPIK 3901, Bavnodde Greensand, east of Bavnodde, between beds 21 and 22. A, dorsal view; B, lateral view.

□2. *Actinocamax verus verus* Miller, MGUH 23718, Bavnodde Greensand, west of Bavnodde, between beds 3 and 4. A, dorsal view; B, lateral view; C, ventral view.

□3. *Actinocamax verus* cf. *antefragilis* Naidin, GPIK 3902, Arnager Limestone Formation, glauconitic, sandy marl at Stampe Å, loc. 5. A, dorsal view; B, lateral view, ×1.5.

□4. *Gonioteuthis praewestfalica* Ernst & Schulz, GPIK 3903, Bavnodde Greensand, Jydegård. A, dorsal view; B, lateral view; C, ventral view; D, view of the anterior end, ×2.

□5. *Gonioteuthis praewestfalica* Ernst & Schulz, GPIK 3904, Bavnodde Greensand, Horsemyre Odde. A; ventral view; B, view of the anterior end, ×2.

□6. *Gonioteuthis westfalica westfalica* (Schlüter), MGUH 23719, Bavnodde Greensand, east of Bavnodde, between beds 24 and 25. A, dorsal view; B, lateral view; C, ventral view; D, dorsal view of the anterior end showing granules, ×3; E, view of the anterior end, ×2.

□7. *Gonioteuthis westfalica westfalica* (Schlüter), MGUH 23720, Bavnodde Greensand, east of Bavnodde, between beds 24 and 25. A, dorsal view; B, view of the anterior end, ×2. This specimen is rather lanceolate in ventral view.

□8. *Gonioteuthis westfalica westfalica* (Schlüter), MGUH 23721, Bavnodde Greensand, west of Bavnodde, between beds 2 and 3. A, ventral view; B, view of the anterior end, ×2. This specimen is rather lanceolate in ventral view.

□9. *Gonioteuthis westfalica westfalica* (Schlüter), MGUH 23722, Bavnodde Greensand, west of Bavnodde, bed 4. A, ventral view; B, view of the anterior end, ×2.

□10. *Gonioteuthis westfalica westfalica* (Schlüter), MGUH 23723, Bavnodde Greensand, west of Bavnodde, between beds 5 and 6. A, dorsal view; B, lateral view of the anterior end showing granules, ×3; C, view of the anterior end, ×2. This specimen is closely similar to *Goniocamax lundgreni* with respect to the depth of the pseudoalvolus (Riedel-Index: 12.2) and the cross-section of the anterior end. It is assigned to *G. w. westfalica*, because it has granules.

□11. *Gonioteuthis westfalica westfalica* (Schlüter), MGUH 23724, Bavnodde Greensand, west of Bavnodde, between beds 5 and 6. A, dorsal view; B, lateral view; C, view of the anterior end, ×2.

□12. *Gonioteuthis ernsti* sp.nov., paratype, MGUH 23725, Bavnodde Greensand, west of Bavnodde, between beds 3 and 5. A, dorsal view; B, lateral view; C, view of the anterior end, ×2.

□13. *Gonioteuthis ernsti* sp.nov., holotype, GPIK 3905, Bavnodde Greensand, west of Bavnodde, between beds 4 and 5. A, dorsal view; B, ventral view; C, view of the anterior end, ×2.

□14. *Goniocamax*(?) sp.nov.(?), MGUH 23731, bottom conglomerate of the Arnager Limestone, west of Arnager. A, dorsal view; B, lateral view; C, ventral view; D, view of the anterior end, ×2.

□15. *Belemnitella propinqua propinqua* (Moberg), holotype, SGU Type 3906, Eriksdal, Scania. A, dorsal view; B, lateral view; C, ventral view; D, view of the split anterior end showing internal characters, ×1.5.

□16. *Belemnitella propinqua propinqua* (Moberg), SGU unregistered, Eriksdal, Scania. A, dorsal view; B, lateral view; C, view of the split anterior end showing internal characters, ×1.5. This specimen was figured as *Actinocamax propinquus*? by Moberg (1885, Pl. 6:22), and later referred to as *Belemnitella mucronata* mut. *anterior* by Stolley (1897, p. 296).

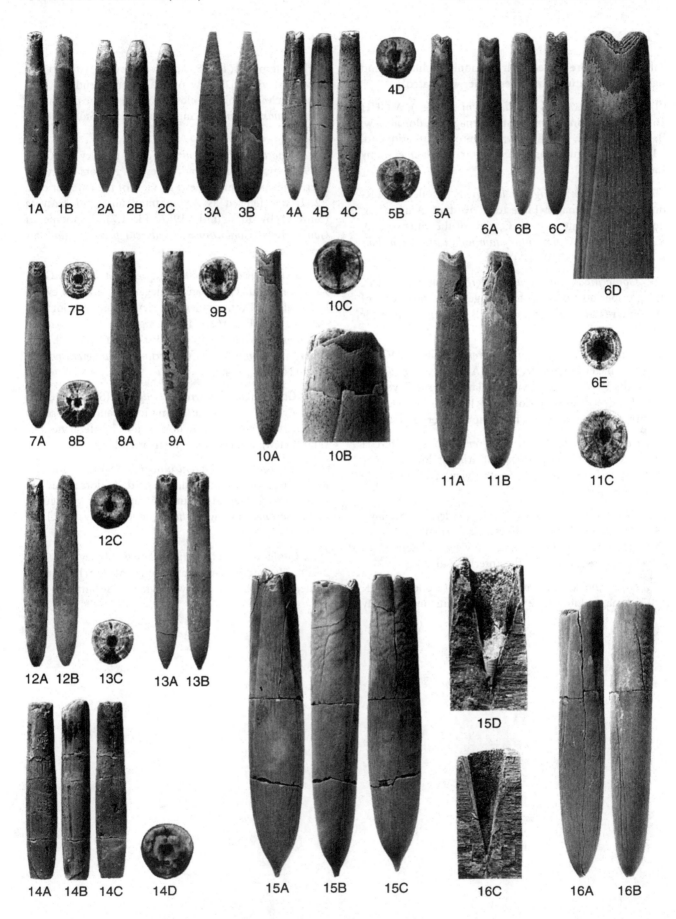

Plate 2

All specimens are coated with ammonium chloride, and figures are of natural size unless otherwise stated.

☐1. *Goniocamax lundgreni lundgreni* (Stolley), MGUH 1690, Arnager Limestone, west of Arnager. A, dorsal view; B, lateral view; C, ventral view; D view of the anterior end, ×1.5. It was figured as *Actinocamax bornholmensis* by Ravn (1918, Pl. 2:7).

☐2. *Goniocamax lundgreni lundgreni* (Stolley), GPIG unregistered, Arnager Limestone, west of Arnager. A, dorsal view; B, lateral view; C, view of the anterior end, ×2. Holotype of *Actinocamax mammillatus* mut. *bornholmensis* Stolley (1897, Pl. 4:1).

☐3. *Goniocamax lundgreni lundgreni* (Stolley), GPIK 3906, uppermost part of the Arnager Limestone, east of Horsemyre Odde. A, ventral view; B, view of the anterior end, ×2.

☐4. *Goniocamax lundgreni lundgreni* (Stolley), MGUH 23726, Arnager Limestone, west of Arnager, 2.0 m ±0.5 m above the base. A, dorsal view; B, lateral view; C, ventral view; D, view of the anterior end, ×1.5. This is the largest specimen of *G. l. lundgreni* from the Arnager Limestone.

☐5. *Goniocamax lundgreni lundgreni* (Stolley), MGUH 23727, Arnager Limestone, west of Arnager, loose. A, dorsal view; B, lateral view; C, ventral view; D, view of the anterior end, ×2.5.

☐6. *Goniocamax* sp.nov.(?) aff. *lundgreni* (Stolley), MGUH 23730, Arnager Limestone Formation, 'Glassmarl' at Muleby Å, loc. 5. A, dorsal view; B, lateral view; C, ventral view; D, view of the anterior end, ×1.5.

☐7. *Goniocamax lundgreni lundgreni* (Stolley), GPIK 3907, Arnager Limestone Formation, glauconitic, sandy marl at Stampe Å, loc. 5. A, dorsal view; B, lateral view; C, ventral view; D, view of the anterior end, ×1.5. A stout specimen which is closely similar to the original of *Actinocamax propinquus* mut. (var.) nov. (Stolley, 1897, Pl. 3:23).

☐8. *Goniocamax lundgreni lundgreni* (Stolley), MGUH 7839, Bavnodde Greensand, Jydegård. A, dorsal view; B, lateral view; C, ventral view; D, view of the anterior end, ×2. It was figured as *Actinocamax lundgreni excavata* (Sinzow) by Birkelund (1957, Pl. 1:7). Lectotype of *Gonioteuthis (Goniocamax) lundgreni postexcavata* Naidin, 1964.

☐9. *Goniocamax lundgreni lundgreni* (Stolley), MGUH 7840, Bavnodde Greensand, Jydegård. A, dorsal view; B, lateral view; C, ventral view; D, view of the anterior end, ×2, (see also Fig. 16A). It was figured as *Actinocamax lundgreni excavata* (Sinzow) by Birkelund (1957, Pl. 1:8). Syntype of *Gonioteuthis (Goniocamax) lundgreni postexcavata* Naidin, 1964.

☐10. *Goniocamax lundgreni lundgreni* (Stolley), GPIK 3908, Arnager Limestone Formation, glauconitic, sandy marl at Stampe Å, loc. 5. A, dorsal view; B, lateral view; C, ventral view; D, view of the anterior end, ×2.

☐11. *Goniocamax lundgreni lundgreni* (Stolley), MGUH 7843, Bavnodde Greensand, Jydegård. A, dorsal view; B, lateral view; C, view of the anterior end, ×2. It was figured as *Actinocamax* aff. *westfalicus* (Schlüter) by Birkelund (1957, Pl. 2:3).

☐12. *Goniocamax lundgreni lundgreni* (Stolley), MGUH 23728, Bavnodde Greensand, east of Forchhammers Odde, bed 53. A, lateral view; B, view of the anterior end, ×2.

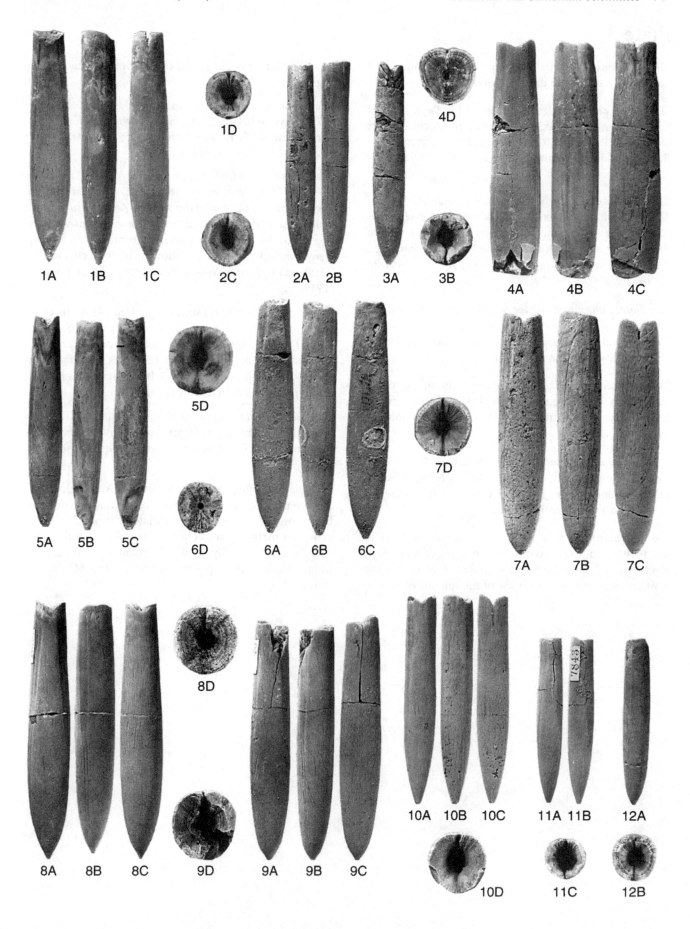

Plate 3

All specimens are coated with ammonium chloride, and figures are natural size unless otherwise stated.

☐1. *Goniocamax lundgreni lundgreni* (Stolley), MGUH 23729, Bavnodde Greensand, between Forchhammers Odde and Skidteper, 40 cm above bed 53. A, dorsal view; B, lateral view; C, ventral view; D, view of the anterior end, ×2.

☐2. *Goniocamax birkelundae* sp.nov., paratype, MGUH 23732, Bavnodde Greensand, Blykobbe Å at Risenholm. A, dorsal view; B, lateral view; C, ventral view; D, view of the anterior end, ×2. It was assigned to *Actinocamax* sp., transitional form between *A. westfalicus* and *A. granulatus* by Birkelund (1957, p. 29).

☐3. *Goniocamax birkelundae* sp.nov., paratype, MGUH 23733, Bavnodde Greensand, east of Forchhammers Odde, bed 53. A, dorsal view; B, lateral view; C, ventral view; D, view of the anterior end, ×2.

☐4. *Goniocamax birkelundae* sp.nov., holotype, GPIK 3909, Bavnodde Greensand, Jydegård. A, dorsal view; B, lateral view; C, ventral view; D, view of the anterior end, ×2.

☐5. *Goniocamax striatus* sp.nov., holotype, GPIK 3915, Bavnodde Greensand, west of Bavnodde, between beds 2 and 4. A, dorsal view; B, lateral view; C, lateral view; D, view of the anterior end, ×2.

☐6. *Goniocamax striatus* sp.nov., paratype, MGUH 23734, Bavnodde Greensand, west of Bavnodde, bed 5. A, dorsal view; B, lateral view; C, ventral view; D, view of the split anterior end, ×1.5; E, view of the split anterior end,

×1.5. The Schatzky distance is ca. 4.7 mm, and the fissure angle is ca. 45°.

☐7. *Goniocamax striatus* sp.nov., paratype, MGUH 23735, Bavnodde Greensand, east of Bavnodde, bed 21. A, dorsal view, ×1.5; B, lateral view, ×1.5; C, ventral view, ×1.5; D, view of the anterior end; ×2.

☐8. *Belemnitella schmidi* sp.nov., paratype, GSC unregistered, *ex* J.A. Jeletzky Collection, Lower Santonian, lower(?) part of the *Inoceramus pachti* Zone *sensu rossico*, Volga area. A, dorsal view; B, lateral view; C, ventral view; D, view of the split guard showing internal characters; E, view of the split anterior end showing internal characters; ×3. It was figured as *Belemnitella propinqua* by Jeletzky (1949, Fig. 1).

☐9. *Belemnitella schmidi* sp.nov., paratype, GSC unregistered, *ex* J.A. Jeletzky Collection. Lower Santonian, lower(?) part of the *Inoceramus pachti* Zone *sensu rossico*, Volga area. A, dorsal view; B, lateral view; C, ventral view; D, view of the anterior end, ×1.5. It was figured as *Belemnitella propinqua* by Jeletzky (1949, Fig. 2).

☐10. *Belemnitella schmidi* sp.nov., holotype, GPIK 3919, Bavnodde Greensand, Jydegård, lower Lower Santonian. A, lateral view; B, ventral view; C, view of the anterior end, ×1.5.

☐11. *Belemnitella propinqua propinqua* (Moberg), GPIK 3921, Bavnodde Greensand, west of Bavnodde, between beds 2 and 3. A, dorsal view; B, lateral view; C, ventral view. It is markedly lanceolate in ventral view and markedly flattened ventrally. Riedel-Quotient is 4.3; Birkelund Index is 4.6; Flattening-Quotient is 1.2; and MLD/LDP Index is 1.3.

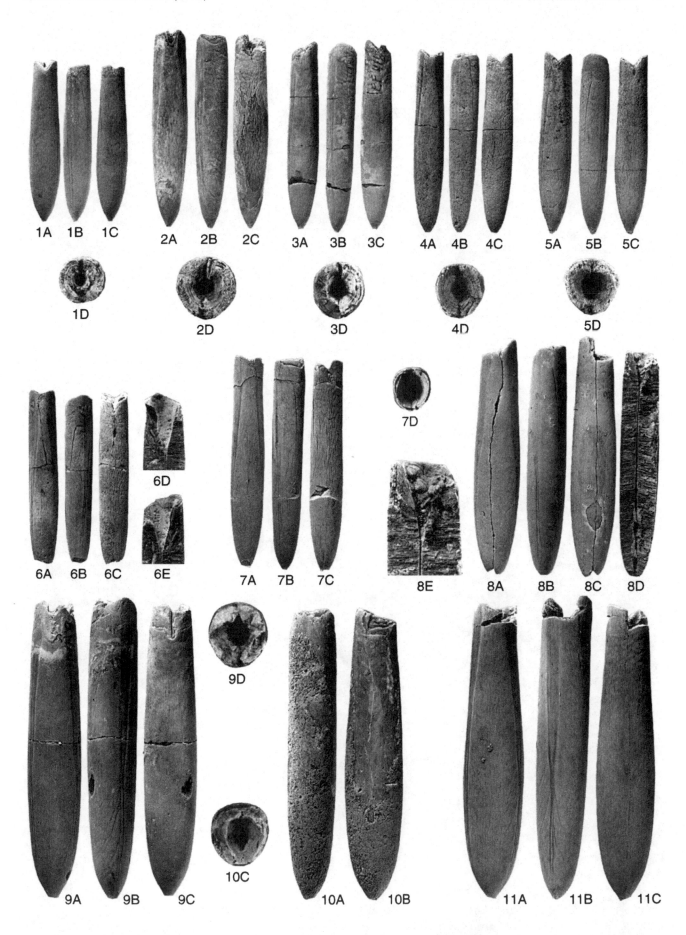

Santonian to Maastrichtian ammonites from Scania, southern Sweden

WILLIAM JAMES KENNEDY AND WALTER KEGEL CHRISTENSEN

Kennedy, W.J. & Christensen, W.K. 1997 08 22: Santonian to Maastrichtian ammonites from Scania, southern Sweden. *Fossils and Strata*, No. 44, pp. 75–128. Oslo. ISSN 0300-9491. ISBN 82-00-37695-8.

Eighteen genera of ammonites are recorded from the Santonian to Maastrichtian rocks of Scania in southern Sweden. Ammonites are fairly common in the Vomb Trough, rare in the Kristianstad Basin and not recorded from the Båstad Basin. The chronostratigraphical assignment of the outcrops on ammonite and belemnite evidence is in agreement. □*Ammonites, Upper Cretaceous, Santonian, Campanian, Maastrichtian, Scania, southern Sweden.*

W. James Kennedy, University Museum Oxford, Parks Road, Oxford OX1 3PW, UK; Walter K. Christensen [wkc@savik.geomus.ku.dk], Geological Museum, University of Copenhagen, Øster Voldgade 5–7, DK–1350 Copenhagen, Denmark; 6th September, 1995; revised 28th August, 1996.

Contents

Introduction

More than a century has passed since Moberg (1885) monographed the Santonian to Maastrichtian ammonites of southern Sweden. He described 21 species and two types of aptychus, the majority of which came from the Vomb Trough.

As early as the 1650s, the Danish polyhistor Ole Worm (1655) figured an ammonite from the Upper Cretaceous of Scania. Stobaeus (1732, Pl. 1:7–9) figured an ammonite as *Cornu Ammonis* (=*Ammonites stobaei* Nilsson, 1827=*Patagiosites stobaei* herein). In the early part of the 19th century, ammonites were dealt with by Wahlenberg (1818) and Nilsson (1826, 1827), and subsequently by Schlüter (1870, 1871–1876), Lundgren (1881), Stolley (1896, 1897, 1930), Hägg (1930, 1935, 1943, 1947, 1954), Lundegren (1933, 1935), Brotzen (1945, 1958), Ødum (1953), Birkelund (1973) and Birkelund & Bromley (1979). Recently, Santonian ammonites from the Köpingsberg-1 borehole in the Vomb Trough were described by Kennedy & Christensen (1993) and those from the uppermost Maastrichtian of the Limhamn pit by Birkelund (1993).

Ammonites are fairly common in the Vomb Trough, rare in the Kristianstad Basin and not recorded from the Båstad Basin. A few ammonites are recorded from the Malmö area. A single specimen referred to *Lewesiceras woodi* Wright, 1979 (p. 312) (=*Pseudopuzosia* sp. of Birkelund, 1973, Pl. 12:1), an Upper Turonian species, is known from Särdal on the Swedish west coast.

Christensen (1973, 1975, 1986, 1993) described the belemnites from the temporary outcrop at Särdal, the Kristianstad and Båstad Basins, in addition to the Vomb Trough, and placed the localities in the international stratigraphical framework on belemnite evidence. Christensen (1985) reviewed the Albian to Maastrichtian of southern Sweden and Bornholm.

The aim of the present paper is to revise the Santonian to Maastrichtian ammonite faunas of Scania in southern Sweden. Most of the specimens described here were collected in the last part of the previous century and the beginning of this century.

Geological setting

The block-faulted Fennoscandian Border Zone acted as a transition zone between the stable Fennoscandian Shield and the subsiding Danish Subbasin during the Permian and Mesozoic (Fig. 1).

Upper Lower and Upper Cretaceous deposits occur in Sweden in the Båstad and Kristianstad Basins northeast of the border zone, in the Vomb Trough within the border zone, and in the Malmö area southwest of the border zone (Fig. 1). The latter area belongs to the Danish Subbasin

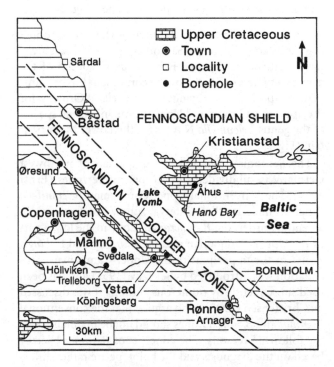

Fig. 1. Map of southern Sweden, showing the distribution of the Upper Cretaceous sedimentary rocks in the Båstad Basin, Kristianstad Basin, Vomb Trough and Malmö area, boreholes and the approximate location of the Fennoscandian Border Zone. Modified from Kennedy & Christensen (1993).

from a tectonic and sedimentological point of view. Upper Cretaceous deposits are also recorded from Särdal on the Swedish west coast north of Båstad, Province of Halland, and from two boreholes in Hanö Bay.

Biocalcarenites and glauconitic sandstones predominate in the basins northeast of the border zone, and the thickness of the strata is less than 250 m. The sedimentary deposits of the Vomb Trough are characterized by detrital clastic sediments – glauconitic, calcareous and clayey siltstones – and the thickness is a little less than 800 m. Limestones, including chalk and marly chalk, prevail in the Malmö area, where the thickness is about 1650 m.

Stratigraphy

Framework. – The Santonian to Maastrichtian of Sweden is of markedly boreal aspect and can be calibrated against the standard successions of northwest Germany, which have been carefully studied during the last forty years by numerous workers (see reviews by Ernst *et al.* 1979; Schulz *et al.* 1984; and Schulz 1996). The Coniacian to Maastrichtian white chalk of Lägerdorf, Kronsmoor and Hemmor, northwest Germany, has been subdivided into 33 faunal zones on the basis of inoceramid bivalves, ammonites, belemnites, echinoids, and crinoids (Schulz *et al.* 1984; Schulz 1996). The faunal zones of the Santonian to lower Lower Maastrichtian provide the stratigraphical framework for the present paper (Fig. 2).

The conventional belemnite zonation. – The conventional belemnite zonation of the Santonian to Lower Maastrichtian of northwest Europe is based upon species of the genus *Gonioteuthis* Bayle for the Santonian–Lower Campanian, species of the genus *Belemnitella* d'Orbigny for the uppermost Lower and Upper Campanian, and species of the genus *Belemnella* Nowak for the Lower Maastrichtian (Fig. 2).

The Santonian and Lower Campanian of northwest Germany is subdivided into 10 belemnite zones (Ernst 1964). The conventional Upper Campanian *Belemnitella* zonation includes three interval zones, and these are, in ascending order, the *B. mucronata* Zone in the lower part of the Upper Campanian and the *B. minor* and *B. langei* Zones in the upper part of the Upper Campanian. This zonation was introduced by Jeletzky (1951) and has been used subsequently by numerous authors. It was discussed by Christensen (1995), who argued that the conventional *B. minor* and *B. langei* Zones should not be maintained for the following reasons. The concepts of *B. minor* Jeletzky based on the diagnosis and the holotype are rather different. In contrast to earlier authors, who interpreted *B. minor* on the basis of the diagnosis, Christensen (1995) interpreted this species with respect to its holotype. In this respect, *B. minor* is a very large species, which was placed

in the *B. mucronata* group owing to its size, surface markings, and internal characters. Moreover, Christensen recognized three chronological subspecies of *B. minor*, which form an evolutionary lineage. *B. minor* I and *B. minor* II Christensen, 1995, predominate in the upper Upper Campanian of Norfolk, whereas *B. minor* III Christensen, 1995, occurs rarely in the lower Lower Maastrichtian.

The concept of *B. langei* Jeletzky is open to question, because it is understood only in the sense of the holotype and paratype. Thus, the concept differs from one author to another. For instance, *B. langei sensu* Birkelund (1957) and *B. langei sensu* Schulz (1978) are not conspecific with *B. langei* of Jeletzky, 1948.

Christensen (1995) offered a *Belemnitella* zonation for the Upper Campanian and lower Lower Maaastrichtian of Norfolk, including five informal zones, in ascending order, the *B. mucronata* and *B. woodi* zones in the lower part of the Upper Campanian, the *B. minor* I and *B. minor* II zones in the upper part of the Upper Campanian, and the *B. minor* III zone in the lower Lower Maastrichtian. The zones were regarded as informal, because the bases of most of them cannot be precisely defined, owing to stratigraphical gaps between the exposures. It was noted that further work is necessary to see if these local zones can be extended to other areas in Europe. Keutgen (1996) has shown subsequently that the zonation recognized in Norfolk is also applicable in northeast Belgium.

Schulz (1979) subdivided the Lower Maastrichtian of northwest Germany into six *Belemnella* zones on the basis of species of two evolutionary lineages of *Belemnella*: the slender subgenus *B. (Belemnella)* Nowak and the stout subgenus *B. (Pachybelemnella)* Schulz.

The subdivision of the Upper Cretaceous of Sweden. – When discussing the subdivision of the Upper Cretaceous of Sweden it should be borne in mind that up to 1930, the zonation was concerned only with the Santonian to Maastrichtian. Cenomanian deposits were first recorded from the Båstad Basin by Lundegren (1932) and Stolley (1932), and the Kristanstad Basin by Christensen (1970). Brotzen (1945) reported Cenomanian, Turonian and Coniacian strata from the Höllviken boreholes.

In the Swedish Santonian to Maastrichtian, the belemnites, above all other macrofossils, have been shown to be of fundamental importance in biostratigraphy and correlation. Christensen (1975, 1986) reviewed the belemnite zonation in great detail, and consequently only a brief outline is given here.

The first reliable subdivision was made by Schlüter (1870), who recognized two belemnite zones, the *Belemnitella mucronata* Zone above and the *Belemnitella subventricosa* Zone (=*Belemnellocamax mammillatus* Zone) below. More belemnite zones were added to Schlüter's stratigraphical scheme by subsequent authors, including Moberg (1885). Stolley (1897, 1930) subdivided the Seno-

Fig. 2. Stratigraphical scheme of the Santonian to lower Lower Maastrichtian, showing faunal zones of northwest Germany, belemnite zones of northwest Europe, zonal belemnites of Balto-Scandia and the age of the localities on belemnite evidence. Sources: column 1, Schulz *et al.* (1984) and Schulz (1996); column 2, Christensen (1986); column 3, Christensen (1975, 1986) and Christensen & Schulz (1997); column 4, Christensen (1975, 1986) and Birkelund & Bromley (1979). The ammonite-bearing beds of Rödmölla–Tosterup, Köpinge and Ignaberga are marked a. The *B. minor* and *B. langei* Zones are placed in square brackets, because Christensen (1995) argued that they should not be maintained. The belemnite zonation of northwest Europe is based upon species of *Gonioteuthis* for the Santonian and Lower Campanian, species of *Belemnitella* for the uppermost Lower and Upper Campanian and species of *Belemnella* for the lower Lower Maastrichtian. The Santonian and Campanian zonal belemnites of Balto-Scandia include species of *Gonioteuthis*, *Goniocamax lundgreni*, *Belemnitella propinqua*, *B. alpha* and *B. mucronata*, as well as the uppermost Lower Campanian *Belemnellocamax mammillatus* and the lower Upper Campanian *Belemnellocamax balsvikensis*.

CHRONO-STRATIGR.	1) FAUNAL ZONES, NW GERMANY	2) BELEMNITE ZONES, NW EUROPE	3) ZONAL BELEMNITES, BALTO-SCANDIA	4) LOCALITIES VOMB TROUGH	KRISTIANSTAD BASIN
LOWER LOWER MAASTRICHTIAN	obtusa	B. obtusa			Bjärnum ? / Åhus ?
	pseudobtusa	B. pseudobtusa			
	lanceolata	B. lanceolata	B. lanceolata		?
UPPER CAMPANIAN upper part	grimmensis/granulosus	[B. langei]			
	langei			Rödmölla-Tosterup	Köpinge
	polyplocum	[B. minor]			
UPPER CAMPANIAN lower part	roemeri		B. mucronata		a
	basiplana/spiniger	B. mucronata		?	Ignaberga / Ivö / Balsberg / Bjärnum
	conica/mucronata		B. balsvikensis/ B. mucronata	?	
LOWER CAMPANIAN upper part	gracilis/mucronata	G. quadrata gracilis/ B. mucronata	D. mammillatus/ B. mucronata/ G. q. scaniensis	a	
	conica/gracilis	G. quadrata gracilis			
	papillosa				
	senonensis			Kullemölla–Lyckås / Kåseberga	Ignaberga
	pilula/senonensis	G. quadrata quadrata U			
LOWER CAMPANIAN lower part	pilula				?
	lingua/quadrata	L			
	granulataquadrata	G. granulata-quadrata	G. granulataquadrata/ B. alpha		a
SANTONIAN U	testudinarius/granulata	G. granulata U	G. granulata	Rödmölla-Tosterup / Eriksdal	?
	socialis/granulata	L			
SANTONIAN M	rogalae/westfalica-granulata	G. westfalica-granulata	G. westfalicagranulata/ B. propinqua	?	
	rogalae/westfalica	G. westfalica U	G. westfalica/ B. propinqua U	?	
SANTONIAN L	coranguinum/westfalica	L	L	?	
	pachti/undulato-plicatus		G. lundgreni/ G. praewestfalica		

nian of Germany and Sweden on the basis of belemnites into, from the bottom to top, the *Westfalicus*-Kreide = Emscher, *Granulaten*-Kreide, *Quadraten*-Kreide = *Mammillaten*-Kreide, and *Mucronaten*-Kreide. This subdivision was used subsequently in Sweden, but translated into *westfalicuskrita*, *granulatuskrita*, *mammillatuskrita*, and *mucronatakrita*.

The German *Quadraten*-Kreide is broadly equivalent to the Lower Campanian. Stolley (1897, 1930) subdivided the *Mucronaten*-Kreide into three zones, mainly on the basis of ammonites. The lower and middle *Mucronaten*-Kreide equate with the Upper Campanian, and the upper *Mucronaten*-Kreide with the Maastrichtian. Stolley correlated the German *Quadraten*-Kreide with the Swedish

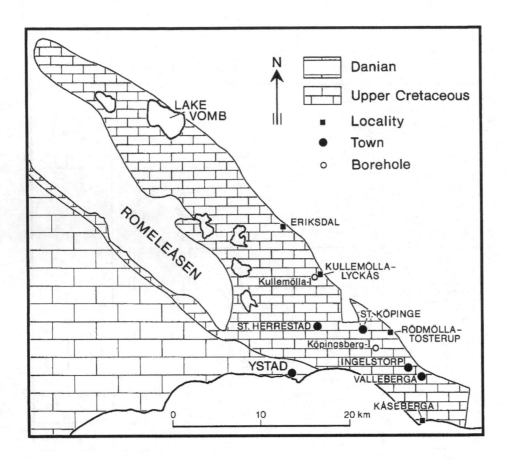

Fig. 3. Map of the Vomb Trough, showing outcrops and boreholes. After Christensen (1986).

Mammillaten-Kreide. Christensen (1975) discussed at length the age of the zone of *Belemnellocamax mammillatus* and concluded that this zone equates with the German *gracilis/mucronata* Zone, the uppermost zone of the Lower Campanian (Fig. 2). Christensen (1986) noted, however, that specimens of *Belemnellocamax balsvikensis*, a species first recognized by Brotzen (1960) and described later by Christensen (1975), were placed in *Belemnellocamax mammillatus* by earlier authors. *B. balsvikensis* occurs in the lower Upper Campanian in beds that can be correlated with the German *conica/mucronata* Zone and the lower part of the overlying *basiplana/spiniger* Zone (Christensen & Schulz 1976). The Swedish *mammillatuskrita*, therefore, spans the uppermost Lower and lower Upper Campanian, and the upper part of the *mammillatuskrita* (zone of *B. balsvikensis*) can be correlated with the lower part of the German *Mucronaten*-Kreide.

   Christensen (1975, 1986, 1993) did not establish formal belemnite zones for the Upper Cretaceous of Sweden, because extensive, continuous sections are not available in the Kristianstad Basin, Båstad Basin and Vomb Trough. He correlated the belemnite faunas of southern Sweden with those of northwest Europe, in particular northwest Germany (Fig. 2) and the Russian Platform.

Locality details

The localities mentioned in the text are listed below. The outcrops of the Kristanstad Basin were described in detail by Christensen (1975) and those of the Vomb Trough by Christensen (1986), who also reviewed earlier work. The stratigraphical age of the localities based on their belemnite and ammonite faunas is discussed below. Most outcrops only expose a few metres of sediments, which can be referred to a specific belemnite zone.

Vomb Trough

Fig. 3

Several informal lithostratigraphical units, including the Eriksdal marl, Kullemölla marl, Lyckås marl, Kåseberga marl, Köpinge sandstone, Valleberga sandstone, and Tosterup conglomerate, have been established for various sedimentary rocks in the trough. Some of these units have been used subsequently as formal formations. They are, however, poorly defined or not defined at all and are therefore used informally here.

Eriksdal. – The abandoned pit at Eriksdal is now covered by scree and overgrown. The Eriksdal marl has yielded a belemnite assemblage consisting of *Gonioteuthis westfalicagranulata*, *Belemnitella propinqua*, and *Actinocamax verus*, which indicate the upper Middle Santonian *G. westfalicagranulata* Zone (Christensen 1986) (Fig. 2). It should be noted, however, that Christensen also reported a few specimens of *Gonioteuthis*, which are closely comparable to *G. westfalica*, possibly indicating that lower middle Santonian strata are also present at Eriksdal.

The following ammonites are here recorded from Eriksdal: *Hauericeras pseudogardeni*, *Nowakites hernensis*, *Glyptoxoceras crispatum*, and *Baculites* ex gr. *capensis* (=*B. brevicosta*? Moberg and *B. incurvatus* Moberg); belemnite and ammonite datings are in agreement.

Kåseberga. – Loose blocks of light, porous, yellow to greyish-yellow Upper Cretaceous marlstone are known from immediately west of the small village of Kåseberga at the sea-shore on the south coast. Christensen (1986) recorded *Belemnitella* ex gr. *alpha/praecursor* and *Gonioteuthis* sp., possibly *G. granulataquadrata* or *G. quadrata*, from the loose blocks, which are thus basal Lower Campanian on belemnite evidence. It is noteworthy, however, that a single specimen of the Cenomanian *Actinocamax* ex gr. *primus/plenus* has been collected recently from a loose block at this locality (Christensen, unpublished). Some of the blocks are therefore Cenomanian.

The ammonites recorded from the blocks are: *Scaphites hippocrepis* III, *Baculites suecicus* Moberg, *B.* ex gr. *capensis* (=*B. incurvatus* Moberg). Belemnite and ammonite datings are in agreement.

Köpinge district. – The Köpinge sandstone, a yellow, highly calcareous, glauconitic sandstone, has been recorded from many small outcrops in the Köpinge district, including Köpingemölla, Svenstorp, Valleberga, Ingelstorp, Herrestad and Fredriksberg. The majority of these outcrops are now inaccessible or very poorly exposed. The Köpinge sandstone was placed in the *Mucronaten*-Kreide or equivalent zones by authors in the last part of the 19th century and the beginning of the 20th century (Schlüter 1870; Moberg 1885; Stolley 1897; Hägg 1930; and others). Lundegren (1933), however, recorded *Belemnellocamax mammillatus* from the Köpinge sandstone, and he therefore assigned the oldest part of the sandstone to the *mammillatuskrita*.

Christensen (1986) recorded two belemnite assemblages from the Köpinge sandstone: (1) an assemblage consisting of *Belemnellocamax mammillatus* and *Belemnitella mucronata*, which is from the uppermost Lower Campanian, and (2) an assemblage of *Belemnitella mucronata* and *B.* aff. *B. langei*, which is from the middle Upper Campanian (Fig. 2). The zone of *Belemnellocamax balsvikensis* and *Belemnitella mucronata* was not recognized,

but this may be due to lack of exposures (Christensen 1986, p. 18).

The following ammonites are recorded: *Gaudryceras* sp., *Pachydiscus haldemsis*, *Patagiosites stobaei*, *Nostoceras junior*, *Lewyites elegans*, *Baculites* cf. *aquilaensis*, *B. angustus*, *B. schlueteri*, *B.* spp., *Trachyscaphites spiniger spiniger* and *Hoploscaphites ikorfatensis*. This is a lower Upper Campanian ammonite association, a dating compatible with the higher belemnite assemblage.

Moberg (1885) recorded the Maastrichtian ammonites *Scaphites constrictus* and *Anagaudryceras* cf. *lueneburgense* from the Köpinge sandstone. These determinations are in error. The record of *S. constrictus* may be based on fragments or microconchs of *Hoploscaphites ikorfatensis*, and that of *A. luenebergense* on *Gaudryceras* sp. (see below).

Kullemölla–Lyckås. – Several closely spaced outcrops are recorded at Kullemölla and Lyckås (see review by Christensen 1986). These are now very poorly exposed or not exposed at all. The Kullemölla and Lyckås marls have yielded a basal Lower Campanian belemnite assemblage of *Gonioteuthis granulataquadrata*, *Belemnitella alpha*, *Actinocamax verus* and *Belemnellocamax* ex gr. *grossouvrei*, indicating the *G. granulataquadrata* Zone (Christensen 1986) (Fig. 2). We record *Hauericeras pseudogardeni* from Lyckås, an occurrence compatible with the belemnite dating.

Rödmölla–Tosterup. – Five outcrops are recorded from Rödmölla–Tosterup (see review by Christensen 1986).

The Campanian part of the sequence consists of alternating calcareous sandstones (Köpinge sandstone) and conglomerates (Tosterup conglomerate). Christensen (1986) recognized two belemnite assemblages from the outcrops. Localities 2 and 4 have yielded an uppermost Lower Campanian belemnite assemblage of *Belemnellocamax mammillatus*, *Belemnitella mucronata* and *Gonioteuthis quadrata scaniensis*?. Loose blocks of the Tosterup conglomerate at loc. 3 yielded the lower Upper Campanian *Belemnellocamax balsvikensis* (Fig. 2).

The following ammonites are recorded from Rödmölla–Tosterup: *Pachydiscus* cf. *subrobustus*, *Patagiosites stobaei*, *Lewyites elegans*, *Trachyscaphites spiniger spiniger*, *Hoploscaphites ikorfatensis* and *Baculites* cf. *aquilaensis*. This is a lower Upper Campanian ammonite association, a dating compatible with the higher belemnite assemblage.

Köpingsberg-1 borehole. – The Santonian ammonites of this borehole were described by Kennedy & Christensen (1993), who recorded *Hauericeras* cf. *pseudogardeni*, *Scalarites* sp., *Baculites* sp. group of *capensis*, *B.* sp. 1, *Boehmoceras krekeleri*, *B. arculus* and *Scaphites kieslingswaldensis fischeri*.

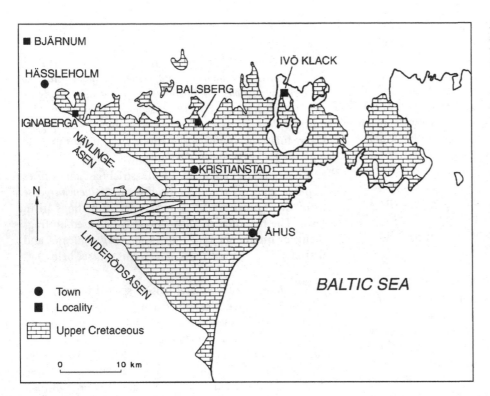

Fig. 4. Map of the Kristianstad Basin, showing the location of the sites dealt with herein. Modified from Siverson (1992).

Kristianstad Basin

Fig. 4

Åhus. – Loose blocks and boulders of the so-called Åhus sandstone are recorded from Revhaken, immediately south of Åhus. The age of the Åhus sandstone has been the subject of much discussion; it is probably from the earliest Maastrichtian but may also be latest Campanian (Christensen 1975). We record *Baculites vertebralis* Moberg *non* Lamarck from the Åhus sandstone. The type material is inadequate, and this taxon is best treated as a *nomen dubium* (see below).

Balsberg. – This locality is a natural cave, which shows biocalcarenites, resting partly on the crystalline basement and partly upon quartz sand. Christensen (1975) recorded the uppermost Lower Campanian *Belemnellocamax mammillatus* and *Belemnitella mucronata* from this outcrop. It should be noted, however, that Christensen (1975, p. 46) also recorded specimens of *Belemnellocamax* with a rather deep pseudoalveolus and many closely spaced conellae. These specimens may be transitional forms between *B. mammillatus* and *B. balsvikensis* or belong to the lower Upper Campanian *B. balsvikensis*. We record '*Baculites vertebralis*' Moberg *non* Lamarck from this outcrop.

Bjärnum. – This abandoned pit showed a 0.5 m thick conglomerate at the base, which rested on weathered crystalline basement. The conglomerate is overlain by about 6 m of calcisiltite with detrital crystalline material and phosphatic nodules.

The uppermost Lower Campanian *Belemnellocamax mammillatus* occurs in the conglomerate. The calcisiltite yielded a basal Lower Maastrichtian belemnite assemblage of *Belemnella lanceolata* and *Belemnitella minor sensu* Christensen (1975), indicating the lower part of *Belemnella lanceolata* Zone *sensu* Schulz (1979). We record *Baculites knorrianus*, a Lower Maastrichtian species, from this locality. Belemnite and ammonite datings are thus in agreement.

Ignaberga new quarry. – In the early 1970s, this pit showed about 10 m of calcarenite with a conglomerate in the middle part of the section (Christensen 1975, Fig. 7). The calcarenites have yielded an uppermost Lower Campanian belemnite assemblage of *Belemnellocamax mammillatus*, *Belemnellocamax* ex gr. *grossouvrei*, *Belemnitella mucronata* and *Gonioteuthis quadrata scaniensis* (Christensen 1975).

In the spring of 1977, downward exploitation revealed a hardground at a horizon not exposed previously (Birkelund & Bromley 1979). A specimen of *Hauericeras* cf. *pseudogardeni* (=*H. pseudogardeni* herein) was recorded from 5 cm below the hardground by Birkelund & Bromley (1979), implying that the calcarenites below the hardground are Upper Santonian – basal Lower Campanian. There is thus a considerable hiatus in the sequence at the hardground, spanning the *G. q. quadrata* and *G. quadrata gracilis* Zones (Fig. 2). Ignaberga new quarry, as well as abandoned pits and mines at Ignaberga, were dealt with in detail by Erlström & Gabrielson (1992).

Ivö Klack (=Blaksudden, Ivöbrottet). – The coarse-grained calcarenites of this pit have yielded an uppermost Lower Campanian belemnite assemblage of *Belemnellocamax mammillatus*, *Belemnellocamax* ex gr. *grossouvrei*, *Belemnitella mucronata* and *Gonioteuthis quadrata scaniensis* (Christensen 1975) (Fig. 2).

We record the following ammonites from Ivö Klack: *Pseudophyllites indra*, *Pachydiscus colligatus*, *Scaphites binodosus*, and *Baculites* spp. Belemnite and ammonite datings are in agreement.

Malmö area
Fig. 1

Limhamn pit. – This abandoned pit lies in the southern outskirts of the town of Malmö. Birkelund (1993) recorded *Scaphites constrictus* from the chalk, which was placed in the *stevensis–chitoniformis* Zone of the brachiopod zonation of Surlyk (1984); this is the uppermost brachiopod zone of the Maastrichtian.

Ulricelund near Näsbyholm. – The abandoned pit at Ulricelund in the southeastern part of the Malmö area was dug in a glacially transported mass of Maastrichtian chalk (Hägg 1954). This chalk yielded the brachiopod *Trigonosemus pulchellus*, which is the index fossil of the *pulchellus–pulchellus* Zone of the brachiopod zonation of Surlyk (1984). This zone is placed in the basal part of the upper Lower Maastrichtian. We record *Scaphites constrictus* from this locality, an occurrence compatible with the brachiopod dating.

Deep boreholes. – Ammonites have been studied from the following deep boreholes: the Höllviken, Svedala and Trelleborg boreholes.

We record *Bostrychoceras polyplocum* (Upper Campanian) from the Trelleborg borehole and *Nowakites hernensis* (Santonian) from the Svedala borehole. The following Upper Campanian to Maastrichtian ammonites are recorded from the Höllviken boreholes: *Saghalinites* sp., *Tragodesmoceras clypeale*, *Hoplitoplacenticeras coesfeldiense*, *Hoploscaphites ikorfatensis*, *H. tenuistriatus* and *Acanthoscaphites* sp.

Systematic palaeontology

Location of specimens. – This is indicated by the following abbreviations: BMNH, Natural History Museum, London; IPB, Paläontologisches Institut der Universität, Bonn; LO, Geological Institute, University of Lund; MGUH, Geological Museum, University of Copenhagen; MNB, Museum für Naturkunde, Berlin; PI, Palaeontological Institute, Uppsala; RM, Naturhistoriska riksmuseet, Stockholm (Swedish Museum of Natural History, Section of Paleozoology); SGU, Sveriges Geologiska Undersökning, Uppsala (Geological Survey of Sweden).

Order Ammonoidea Zittel, 1884

Suborder Lytoceratina Hyatt, 1889

Superfamily Tetragonitaceae Hyatt, 1900

Family Gaudryceratidae Spath, 1927

Genus *Gaudryceras* de Grossouvre, 1894

Type species. – *Ammonites mitis* Hauer, 1866, by subsequent designation of Boule *et al.* (1906, p. 183[11]).

Gaudryceras sp.
Fig. 5A–B

Description. – The specimen is a well-preserved fragment of body chamber from Köpinge, with calcite-replaced shell preserved. The maximum preserved whorl height is 46 mm; the whorl breadth to height ratio is 0.76, the whorl section ovoid, with feebly convex inner flanks, flattened, convergent outer flanks, broadly rounded ventrolateral shoulders and a feebly convex venter.

Ornament is of very fine lirae, convex across the inner flank, concave on the outer flank, strongly projected on the outer ventrolateral shoulder and broadly convex across the venter. There are periodic scale-like ridges and shallow grooves.

Discussion. – This fragment is specifically indeterminate and might even be referred to *Anagaudryceras* Shimizu, 1934. Moberg (1885) recorded *Anagaudryceras* cf. *lueneburgense* (Schlüter, 1872) from the Köpinge sandstone. This is a Maastrichtian species; the present specimen may explain the record as being based on one of the long-ranging, finely ribbed *Gaudryceras* species.

Occurrence. – Köpinge sandstone, Köpinge, probably middle Upper Campanian.

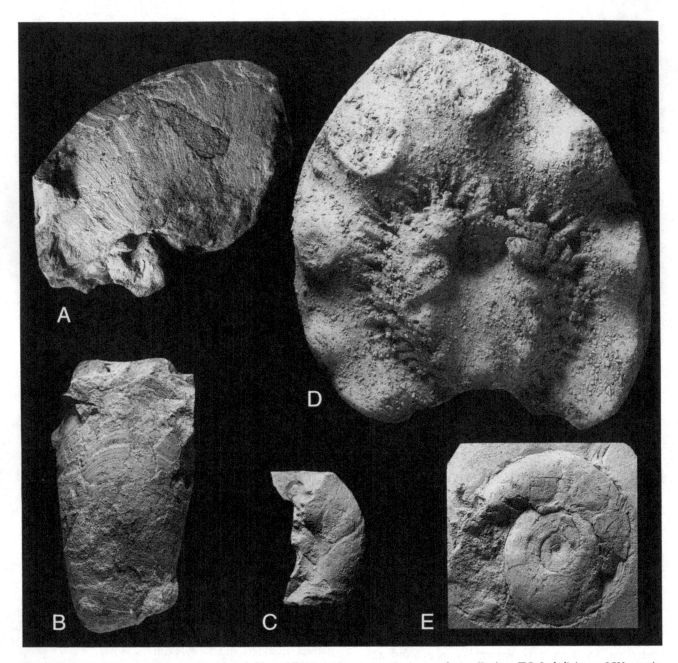

Fig. 5. □A–B. *Gaudryceras* sp. SGU unregistered, probably middle Upper Campanian, Köpinge sandstone, Köpinge. □C. *Saghalinites* sp. SGU unregistered, Lower Maastrichtian, Höllviken-1 borehole, 540.65–540.75 m. □D. *Pseudophyllites indra* (Forbes, 1846). SGU unregistered, uppermost Lower Campanian of Ivö Klack. □E. *Hauericeras (Hauericeras) pseudogardeni* (Schlüter, 1872). SGU Type 3875, the original of *Ammonites* n.sp.? of Moberg (1885, Pl. 2:7), from the upper Middle Santonian of Eriksdal. All ×1.

Family Tetragonitidae Hyatt, 1900

Genus *Saghalinites* Wright & Matsumoto, 1954

Type species. – *Ammonites cala* Forbes, 1846, p. 104, Pl. 87:4, by original designation of Wright & Matsumoto (1954, p. 110).

Saghalinites sp.

Fig. 5C

Description. – The specimen is a crushed composite mould of 120° of phragmocone with a maximum preserved whorl height of 15.5 mm. The shell surface is smooth, but for widely spaced prorsiradiate constrictions, convex on the inner flank, concave on the outer flank and feebly convex across the venter.

Discussion. – This fragment is specifically indeterminate, but may belong in *Saghalinites wrighti* Birkelund, 1965 (see revisions in Birkelund 1993, p. 45, Pls. 1:6–7; 2:1, 3–4, 6, and Ward & Kennedy 1993, p. 21, Figs. 19.3–19.4, 19.8, 19.12, 20.1–20.3), a species that ranges from upper Lower to lower Upper Maastrichtian.

Occurrence. – Lower Maastrichtian, Höllviken-1 borehole, 540.65–540.75 m.

Genus *Pseudophyllites* Kossmat, 1895

Type species. – *Ammonites indra* Forbes, 1846, p. 105, Pl. 11:7, by original designation of Kossmat (1895, p. 137[41]).

Pseudophyllites indra (Forbes, 1846)

Fig. 5D

Synonymy. – □1846 *Ammonites Indra* – Forbes, p. 105, Pl. 11:7. □1992 *Pseudophyllites indra* (Forbes) – Kennedy & Henderson, p. 398, Pls. 3:7–9, 13–27; 4:1–3 (with full synonymy). □1993 *Pseudophyllites indra* (Forbes) – Ward & Kennedy, p. 22, Figs. 17.8, 18.9–10, 19.7, 19.9, 19.13, 19.21, 21.2, 22.1, 27.6. □1993 *Pseudophyllites indra* (Forbes) – Hancock & Kennedy, p. 153, Pl. 1:3–4. □1993 *Pseudophyllites indra* (Forbes) – Kennedy & Hancock, p. 577, Pl. 1:4, 7.

Types. – Lectotype, designated by Kennedy & Klinger (1977, p. 182), is BMNH C51068, the original of Forbes (1846, Pl. 11:7). The paralectotypes are C51069–73, all from the Upper Maastrichtian Valudavur Group of Pondicherry, south India.

Description. – The specimen, from Ivö Klack, is a worn internal mould of a single chamber with a maximum preserved whorl height of 120 mm. The whorl section is compressed oval, with a massive septal lobe.

Discussion. – The massive septal lobe alone serves to identify this fragment as *Pseudophyllites indra*, rather than *Pseudophyllites loryi* (Kilian & Reboul, 1909, p. 18, Pl. 1:4–5) and its synonyms *P. latus* (Marshall, 1926, p. 149, Pls. 20:6, 6d; 32:1–2), and *P. peregrinus* Spath, 1953, p. 7, Pl. 1:6–9), as discussed by Henderson & McNamara (1985, p. 50).

Occurrence. – *Pseudophyllites* may first appear in the Upper Santonian and ranges with certainty from Lower Campanian to uppermost Maastrichtian. *P. indra* is known from Sweden, Northern Ireland, southwestern France, Spain, Poland, Austria, New Jersey, The U.S. Gulf Coast, British Columbia, Alaska, Saghalin, Japan, Chile, Brazil, western Australia, New Zealand, South India, Madagascar, Zululand (South Africa).

Suborder Ammonitina Hyatt, 1889

Superfamily Desmocerataceae Zittel, 1895

Family Desmoceratidae Zittel, 1895

Subfamily Puzosiinae Spath, 1922

Genus and subgenus *Hauericeras* de Grossouvre, 1894

(=*Schlueteria* Rollier, 1922, p. 359, *non* Fritsch *in* Fritsch & Kafka, 1887, p. 33; *Pseudogardenia* Tomlin, 1930, p. 23)

Discussion. – Matsumoto *in* Matsumoto, Toshimitsu & Kawashita (1990) distinguished two subgenera in *Hauericeras*, the nominotypical subgenus, with feeble ventrolateral riblets/nodes, and *Gardeniceras* Matsumoto & Obata, 1955 (p. 134) (type species *Ammonites gardeni* Baily, 1855, p. 450, Pl. 11:3), which lacks riblets/nodes. He also concluded that Hauericeratinae Matsumoto, 1938, should be incorporated into Puzosiinae, a view accepted here.

Hauericeras (*Hauericeras*) *pseudogardeni* (Schlüter, 1872)

Figs. 5E, 6

Synonymy. – □1872 *Ammonites pseudo-Gardeni* – Schlüter, p. 54, Pl. 16:3–6. □1875 *Haploceras pseudogardeni* Schlüter – Neumayr, p. 915. □1885 *Ammonites* n.sp.? – Moberg, p. 25, Pl. 2:7. □1894 *Hauericeras pseudo-Gardeni* Schlüter – De Grossouvre, p. 220, Text-fig. 81. □ *non* 1898 *Hauericeras pseudo-gardeni* Schlüter – Mariani, p. 8 (1), Fig. 6. □1899 *Ammonites pseudogardeni* Schlüter var. *nodatus* – Schlüter, p. 411. □1899 *Ammonites pseudo-gardeni* Schlüt. – Schlüter, p. 411. □1905 *Hauericeras pseudogardeni* Schlüter – Wegner, p. 207. □1905 *Hauericeras buszii* – Wegner, p. 208, Pl. 8:1a–b. □1905 *Hauericeras buszii* var. *nodata* – Wegner, p. 209, Pl. 8:1a. □1905 *Hauericeras buszii* var. *costata* – Wegner, p. 208, Pl. 8:1b. □1906 *Hauericeras pseudo-Gardeni* Schlüter – Müller & Wollemann, p. 14, Pls. 4:1–4; 8:3. □1920 *Hauericeras pseudogardeni* Schlüter – Köplitz, p. 64, Pl. 8:24. □1920 *Hauericeras buszii* Wegner – Köplitz, p. 64. □1925 *Hauericeras Pseudo-Gardeni* Schlüter – Diener, p. 95. □1930 *Hauericeras* cf. *pseudo-Gardeni* Schlüter – Hägg, p. 60. □1931 *Hauericeras pseudogardeni* Schlüt – Riedel, p. 694, Pl. 78:7. □1931 *Hauericeras buszii* Wegner – Riedel, p.

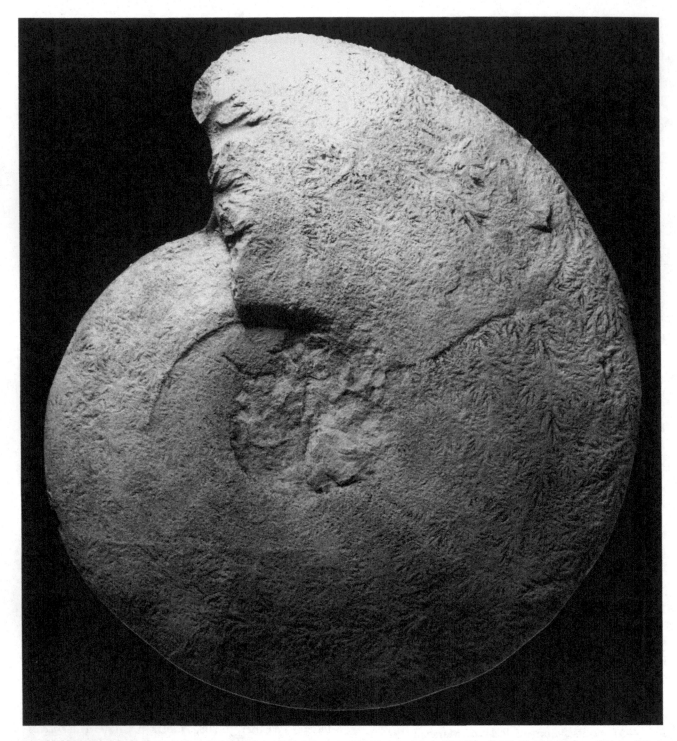

Fig. 6. Hauericeras (Hauericeras) pseudogardeni (Schlüter, 1872). Lectotype, IPB 48, the original of Schlüter (1872, Pl. 16:5–6), from Dülmen, Westphalia. ×0.75.

695. □1935 *Hauericeras pseudo-Gardeni* (Schlüter) – Hägg, p. 59. □1938 *Hauericeras pseudogardeni* Schlüt. – Roman, p. 406, Pl. 41:393. □?*non* 1951 *Hauericeras* cf. *pseudogardeni* Schlüter – Mikhailov, p. 81, Pl. 12:50. □?1953 *Hauericeras pseudo-gardeni* Schlüt. – Ødum, p.

23, Pl. 4:11. □1957 *Hauericeras pseudogardeni* (Schlüter) – Wright, p. L371, Fig. 485:1a–1d. □1964 *Hauericeras pseudogardeni* – Arnold, p. 12. □1979 *Hauericeras* cf. *pseudogardeni* (Schlüter) – Birklund & Bromley, p. 173, Fig. 3. □1987 *Hauericeras pseudogardeni* (Schlüter) –

Kennedy & Summesberger, p. 28. □1990 *Hauericeras pseudogardeni* (Schlüter) – Matsumoto *in* Matsumoto, Toshimitsu & Kawashita, p. 439, Fig. 1. □1993 *Hauericeras* cf. *pseudogardeni* (Schlüter) – Kennedy & Christensen, p. 152, Fig. 2a.

Type. – Lectotype, by the subsequent designation of Matsumoto *in* Matsumoto *et al.* (1990, p. 440), is IPB 48, the original of Schlüter (1872, Pl. 16:5–6), from Dülmen, Westphalia, Germany (Fig. 6).

Description and discussion. – Moberg (1885, p. 25, Pl. 2:7) figured as *Ammonites* n.sp. a juvenile of this species 46 mm in diameter from Eriksdal (SGU Type 3875; Fig. 5E herein). A second specimen described by him as *Ammonites* sp. (1885, p. 25, Pl. 2:6) from Eriksdal was referred to *Ammonites clypealis* of Schlüter by Hägg (1930), but seems rather to be a *Nowakites hernensis* (Schlüter, 1867), as discussed below. Birkelund & Bromley (1979, p. 173, Fig. 3) described as *H.* cf. *pseudogardeni* a phragmocone of a juvenile from Ignaberga that also belongs here (PI Sk 70), as follows: 'The fully septate nucleus is 85 mm in diameter. It is rather involute, the umbilical ratio being about 22%. The venter is sharp and high and the sides are flat. The umbilical shoulder is not preserved. The specimen carries six or seven constrictions on the outermost whorl preserved. They are deep and slightly sigmoid. The sutures are fairly well preserved and extremely incised.

In spite of the poor preservation, the specimen can be referred to the genus *Hauericeras* on the basis of the cross-section, constrictions and suture lines. Its closest relative seems to be the type of the genus, *Hauericeras pseudogardeni* (Schlüter, 1872). Thus, a comparison with material from Dülmen, Westphalen, kept in Münster Museum, and figured material from that region (Schlüter 1872; Müller & Wollemann 1906) shows good agreement with the specimen. The only difference lies in the constrictions, which may be slightly more curved in our specimen. The closest comparable specimen is figured by Müller & Wollemann (1906, Pl. 4:4).'

We also examined six specimens from Braunschweig in the Institut für Paläontologie, Bonn (unregistered), and four in the collection of the Muséum National d'Histoire Naturelle, Paris (MNHP R51769–51772).

The early growth stages are shown by the Eriksdal specimen and the specimens from Braunschweig, all of which are preserved as crushed composite moulds. Coiling is involute, the umbilicus small (17–20% of diameter), shallow, with a low, flattened wall and sharp umbilical shoulder. The whorl section is very compressed, lanceolate, with the greatest breadth below mid-flank; the degree of compression has been greatly accentuated by *post-mortem* compaction. The moulds show very clearly that the phragmocones had a solid, sharp keel, demarcated from the venter by a groove. Specimens may be smooth to a diameter of 60 mm or may have delicate umbilical and ventral ribs. One specimen shows delicate biconcave constrictions, four per half whorl, at a diameter of 57 mm, together with indications of delicate outer flank ribs. Beyond 60 mm diameter, ornament is much more marked, with delicate biconcave prorsiradiate ribs that increase by branching and intercalation, strengthening markedly on the ventolateral shoulder, where they project strongly forwards, and may strengthen into crescentic outer lateral–ventrolateral nodes on some but not all ribs. MNHP R51769 shows a combination of constrictions, outer flank ribs and tubercles extending to a phragmocone diameter of 130 mm; MNHP R51770 is, in contrast, virtually smooth at the same diameter. What may be a small adult with ribbed, nodate and constricted phragmocone has half a whorl of body chamber, with three narrow, biconvex constrictions with associated low broad collar ribs, of which the adapical rib bears a small bullate to feebly crescentic outer lateral node. There are traces of delicate riblets and striae on the interspaces between, but no tubercles. A second small adult is 155 mm in diameter; it has half a whorl of body chamber preserved. The phragmocone is again ribbed and constricted, with numerous delicate outer lateral nodes. On the body chamber, there are five biconcave prorsiradiate constrictions, again associated with collar-ribs, the adapical strengthened into a crescentic outer lateral tubercle. The interspaces are nearly smooth. MNHP 51772 shows comparable body chamber characteristics at a much smaller size. Although the shape is distorted into an ellipse, the major diameter is only 110 mm. If the last three specimens are correctly interpreted as adult *H.* (*H.*) *pseudogardeni*, they may be microconchs of the species.

The lectotype (Fig. 6) is an internal mould, still septate at 237 mm. Coiling is involute, with 62% of the previous whorl covered, the umbilicus small (24% of the diameter), shallow, with a flattened-outward-inclined umbilical wall and narrowly rounded, sharp umbilical shoulder. The whorl section is compressed, lanceolate, with the greatest breadth around mid-flank. A pronounced facet is associated with a break in the even profile of the whorl section on the outer flank; the venter is acute. The outer flank region between facet and venter bear scarcely discernible traces of delicate, markedly prorsiradiate ribs on the adapical 90° sector of the outer whorl. There are traces of three constrictions on the adapertural 120° of the outer whorl, 60° apart. They are narrow, shallow, straight and prorsiradiate on the innermost flank, then flexed and feebly concave on the inner flank, feebly convex across the middle of the flank, more markedly concave on the outer, and projected strongly forward on the outermost flank. A further large, wholly septate specimen from Dülmen in the Bonn collections is 290 mm in diameter and retains calcite-replaced shell over much of its surface. There are indications of a further 240° of umbilical seam of a now missing outer whorl to an umbilical diameter of 115 mm, corresponding to an estimated diameter of 440 mm. Coiling is a little more evolute than in the previous specimen, the umbilicus comprising 26% of the diameter, with 51% of the previous whorl covered, the umbilical wall low, flattened, outward-inclined, with a sharp umbilical shoulder. The whorl section is compressed, lanceolate, with a whorl breadth to height ratio of 0.46, the greatest breadth below mid-flank. The venter is blunt on the mould but sharp and acute where replaced shell is present, the keel separated from the venter by a marked groove. There are nine constrictions on the outer whorl, narrow, moderately deep, and preceded by a blunt, coarse collar-rib. Ribs and constriction are prorsiradiate, feebly concave across the inner flank, feebly convex across mid-flank and feebly concave on the outer flank, sweeping forwards and projected markedly across the ventrolateral shoulder to form an acute chevron across the siphonal keel. The shell surface between constrictions is variably preserved, but on the first half whorl it bears low irregular ribs and striae of variable length and strength. They are weak, straight and prorsiradiate on the inner flank, but strengthen, sweep forwards and are markedly prorsiradiate on the outer flank and ventrolateral shoulder. These large discs are interpreted as macroconch phragmocones.

Interpretation of the Braunschweig and Dülmen specimens as a probable macroconch and microconch pair must be regarded as tentative. From the material available it seems that internal moulds may be virtually smooth (but for constrictions), whereas the shell surface and composite moulds are much more markedly ribbed. But even some composite moulds are virtually smooth, although this might be due to *post-mortem* effects. If this interpretation is accepted, the variety *nodatum* of Schlüter (1899) and *Hauericeras buszii* Wegner, 1905 (p. 209, Pl. 8:1a–b), are synonyms of *pseudogardeni*. The best previous illustrations of the species are those of Müller & Wollemann (1906) of material from Braunschweig, which included constricted ribbed/nodate phragmocones up to 160 mm diameter (1906, Pls. 4:1; 8:3) as well as smooth, delicately constricted juveniles. The *Hauericeras* cf. *pseudogardeni* of Mikhailov (1951, p. 81, Pl. 12:50) has concave constrictions and seems far too evolute to be referred to the present species.

Matsumoto *in* Matsumoto *et al.* (1990, p. 451) thought *Ammonites mengedensis* Schlüter, 1876 (p. 154, Pl. 40:9) might be the microconch of *H. (H.) pseudogardeni*, but this is a significantly older species (Kaplan & Kennedy 1994) and does not co-occur with *H. (H.) pseudogardeni* in any of the collections we have studied.

Hauericeras (H.) antiquum Collignon (1961, p. 75, Fig. 12) from the Lower Coniacian of Madagascar appears to represent the stock ancestral to *H. (H.) pseudogardeni*, having the shell shape of *Hauericeras* plus delicate ventral ribs, but no constrictions. It is transitional to the Middle Turonian *Puzosia (Puzosia) serratocarinata* Kennedy &

Cobban 1988 (p. 595, Figs. 2; 4:1–3), from northern Mexico, with a fastigate venter but no well-differentiated keel as is present in *Hauericeras*.

Occurrence. – Middle Santonian to Lower Campanian of the Münster Basin and Braunschweig, Germany; Lyckås, Eriksdal and Ignaberga, southern Sweden, plus borehole records from Köpingsberg (Kennedy & Christensen 1993) and Höllviken (Ødum 1953).

Family Pachydiscidae Spath, 1922

Genus and subgenus *Pachydiscus* Zittel, 1884

(=*Parapachydiscus* Hyatt, 1900, p. 570; *Joaquinites* Anderson, 1958, p. 218; *Pseudomenuites* Matsumoto, 1955, p. 169).

Type species. – *Ammonites neubergicus* Hauer, 1858, p. 12, Pl. 2:1–4, by subsequent designation of de Grossouvre (1894, p. 177).

Pachydiscus (Pachydiscus) colligatus (Binkhorst, 1861)

Figs. 7–10

Synonymy. – □1861 *Ammonites colligatus*, Nobis – Binkhorst, p. 25 (*pars*) Pl. 8 only. □1986a *Pachydiscus (Pachydiscus) colligatus* (Binkhorst) – Kennedy, p. 36, Figs. 13–14. □1987 *Pachydiscus (Pachydiscus) colligatus* (Binkhorst) – Kennedy, p. 162, Pls. 1:1–2; 2:1–2; 3; 4:4–5 (with full synonymy). □1993 *Pachydiscus (Pachydiscus)* cf. *colligatus* (Binkhorst) – Hancock & Kennedy, p. 162.

Types. – Lectotype, by the subsequent designation of Kennedy (1987, p. 162), is the original of Binkhorst (1861, Pl. 8), MNB unregistered, from the lower Upper Campanian of Jauche, Brabant, Belgium (Kennedy 1987, Pls. 1:1–2; 2:1–2). The paralectotypes belong to several different species as discussed by Kennedy (1987, pp. 162–163).

Description. – Internal moulds of the earliest growth stages seen have whorl heights of 22 and 40 mm (Fig. 7A–B, E–F). At this size, coiling is moderately involute, the umbilicus small, deep, with a feebly convex wall and broadly rounded umbilical shoulder. The whorl section is depressed reniform, with the greatest breadth just outside the umbilical shoulder and whorl breadth to height ratios of up to 1.31. Low, blunt, narrow primary ribs arise at the umbilical seam and strengthen into feeble to incipient bullae on the umbilical shoulder. These give rise to one or two ribs, with single intercalated ribs between. Ribs are straight and prorsiradiate on the inner flank, then flexing forwards and concave on the outer flank and ventrolateral

Fig. 7. Pachydiscus (Pachydiscus) colligatus (Binkhorst, 1861), uppermost Lower Campanian, Ivö Klack. □A–B. SGU unregistered. □C. SGU unregistered. □D. Silicone squeeze from an external mould of a juvenile, SGU unregistered. □E–F. SGU unregistered. All ×1.

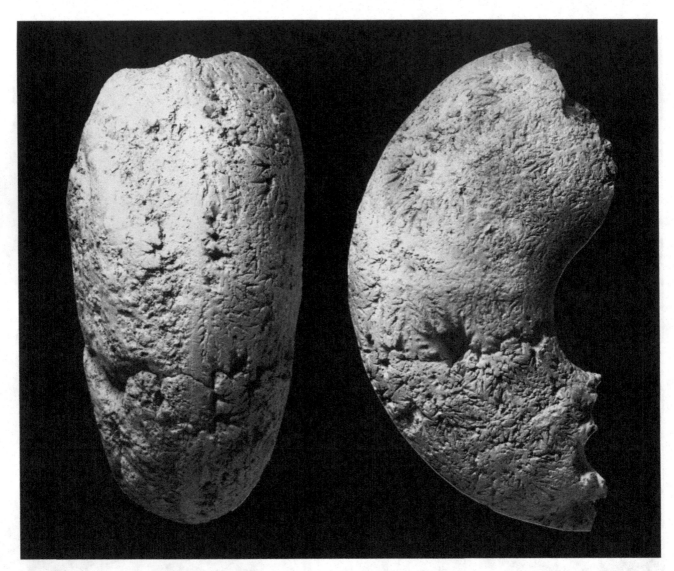

Fig. 8. Pachydiscus (Pachydiscus) colligatus (Binkhorst, 1861), uppermost Lower Campanian, Ivö Klack. SGU unregistered. All ×1.

shoulder where they weaken, and cross the venter in a very shallow convexity. A partial external mould of a juvenile with an estimated diameter of 77 mm shows the ornament of the outer shell surface to have been much sharper (Fig. 7D). Eleven ribs per whorl arise at the umbilical seam, and strengthen across the umbilical wall, most developing into small, sharp bullae. These give rise to pairs of narrow sharp ribs, with single intercalated ribs between. In some larger moulds (Fig. 8), this alternation of primary and intercalated ribs extends to whorl heights of up to 70 mm. Beyond this, inner flank ornament weakens. Fig. 7C shows the ornament of the inner whorls of a specimen with replaced shell at a diameter of 140 mm. Inner flank ornament is very weak. There are no umbilical bullae, and very weak riblets are prorsiradiate and distant

on the inner half of the flank, strengthening only on the outer flank and venter, which they cross in a broad convexity. The ribs are subdued, and separated by very broad interspaces, as on the only sector of the lectotype where the outer shell surface is preserved (Kennedy 1987, Pl. 1:1). Beyond this diameter, the unfigured side of this specimen, which is septate to the maximum preserved diameter of 210 mm, has only obscure distant low flank ribs on the last half whorl (Figs. 9–10). Other, larger specimens show such distant ribs extending up to 330 mm diameter.

Discussion. – P. (P.) colligatus from Ivö Klack is represented by numerous specimens in museum collections. The species was revised and discussed by Kennedy (1987),

Fig. 9. Pachydiscus (Pachydiscus) colligatus (Binkhorst, 1861), uppermost Lower Campanian, Ivö Klack. SGU unregistered (see also Fig. 10). ×0.85.

who illustrated the type material, and discussed differences from other closely allied species.

Occurrence. – The lectotype is from the lower Upper Campanian of Jauche, Brabant, Belgium. In Sweden the species occurs in the uppermost Lower Campanian rocks at Ivö Klack. It is also known from the Upper Campanian of Royan (Charente-Maritime), possibly Tercis (Landes), France, and, perhaps, Madagascar.

Fig. 10. Pachydiscus (Pachydiscus) colligatus (Binkhorst, 1861), uppermost Lower Campanian, Ivö Klack. SGU unregistered (see also Fig. 9). ×0.8.

Pachydiscus (Pachydiscus) haldemsis (Schlüter, 1867)

Fig. 11

Synonymy. – □1867 *Ammonites haldemsis* – Schlüter, p. 19, Pl. 3:1. □1872 *Ammonites haldemsis* – Schlüter, p. 70. □1872 *Ammonites* cf. *auritocostatus* – Schlüter, p. 70, Pl. 22:6–7. □1872 *Ammonites Galicianus* Favre – Schlüter, p. 63, Pl. 19:3–5, Pl. 20:9. □1885 *Ammonites Oldhami* Sharpe – Moberg, p. 23, Pl. 3:1. □1889 *Pachydiscus galicianus* Favre – Griepenkerl, p. 101. □1894 *Pachydiscus koeneni* – de Grossouvre, p. 178. □1902 *Pachydiscus galicianus* (Schlüter) – Wollemann, p. 103. □1913 *Pachydiscus oldhami* Sharpe – Nowak, p. 362, Pls. 41:16; 43:31; 45:43. □?1913 *Pachydiscus kaliszanensis* – Nowak, p. 359, Pl. 40:8. □1913 *Pachydiscus haldemensis* Schlüter – Nowak, p. 349. □1925 *Pachydiscus koeneni* Grossouvre – Diener, p. 106. □1935 *Pachydiscus* ex. aff. *galicianum*

Fig. 11. Pachydiscus (Pachydiscus) haldemsis (Schlüter, 1867). □A–B. LO 721t, the original of *Ammonites oldhami* Sharpe of Moberg (1885, Pl. 3:1), probably middle Upper Campanian of Köpinge. □C. Macroconch phragmocone, IPB 50b, the original of Schlüter (1872, Pl. 19:3–4), from Haldem, Westphalia, Germany. All ×1.

Favre – Brinkmann, p. 5. □1951 *Pachydiscus* cf. *Koeneni* Grossouvre – Mikhailov, p. 60, Pl. 10:47. □1951 *Pachydiscus oldhami* (Sharpe) – Wright & Wright, p. 20 (*pars*). □1954 *Pachydiscus (Parapachydiscus) oldhami* (Sharpe) – Hägg, p. 57. □1955 *Pachydiscus haldemensis* Schlüter – Matsumoto, p. 168. □1955 *Pachydiscus ambiguus* Grossouvre – Matsumoto, p. 169. □1957 *Pseudomenuites ambiguus* (de Grossouvre) – Wright, p. 380. □1959 *Pachydiscus koeneni* Grossouvre – Naidin & Shimanskij, p. 185, Pl. 9:1. □1964 *Pachydiscus koeneni* Grossouvre – Giers, p. 263, Pl. 5:1, *non* 2; *non* Fig. 5. □1974 *Menuites ambiguus* (Grossouvre) – Naidin, p. 182, Pl. 63:2. □1974 *Pachydiscus koeneni* Grossouvre, 1894 – Naidin, p. 186, Pl. 65:2–3. □1980 *Pachydiscus koeneni* Grossouvre – Błaskiewicz, p. 42, Pls. 26:1–2; 27:1–4; 28:1–4; 34:3–4. □1984 *Pachydiscus (Pachydiscus) haldemsis* (Schlüter) – Kennedy & Summesberger, p. 158, Pls. 4:1–5; 5:1; 6:2; 7:1–11; 13:1 (with additional synonymy). □1986a *Pachydiscus (Pachydiscus) haldemsis* (Schlüter) – Kennedy, p. 45, Pls. 4:1–3; 5:7–14; Figs. 11A–D, F–G, 17.

Types. – The lectotype of *Ammonites haldemsis* Schlüter, 1867 (p. 19), by subsequent designation of Kennedy & Summesberger (1984, p. 158), is an unregistered specimen in the Schlüter Collection (IPB), the original of Schlüter (1867, Pl. 3:1), refigured by Kennedy (1986a) as Fig. 11A–B. The lectotype of *P. koeneni* de Grossouvre, 1894 (p. 178) is in the same collection and was refigured by Kennedy (1986a) as Fig. 11D. Both are from the Upper Campanian of Haldem, Westphalia.

Description. – The original of *Ammonites Oldhami* Sharpe of Moberg (1885, Pl. 3:1; Fig. 11A–B herein), from Köpinge, is LO 721t. It is a somewhat worn and crushed phragmocone with the following dimensions: D=86.0 (100); Wb=26.3 (30.6); Wh=38.4 (44.7); Wb:Wh=0.68; U=25.0 (34.4). Coiling is evolute, umbilicus shallow, with outward-inclined wall and broadly rounded shoulder. Whorl section is compressed with whorl breadth to height ratio 0.68, the greatest breadth below mid-flank. Flanks are feebly convex, ventrolateral shoulders and venter broadly rounded. Primary ribs arise at the umbilical seam and strengthen across the umbilical wall and shoulder. They are straight and prorsiradiate on the inner to middle flank, flexing forwards and feebly concave on the outermost flank and ventrolateral shoulder, and feebly convex on the venter. One or two intercalated ribs arise either low or high on the flank and strengthen to match the primary ribs on the outermost flank, ventrolateral shoulders and venter. Suture is typical for genus.

Discussion. – The lectotype of *Ammonites haldemsis* Schlüter, 1867, is a microconch, of which *Pachydiscus koeneni* de Grossouvre, 1894, is the macroconch, as demonstrated by Kennedy & Summesberger (1984). The present specimen is a small, juvenile macroconch. P. (P.)

koeneni most closely resembles *P. (P.) oldhami* (Sharpe, 1855) (p. 32, Pl. 14:2; see revision in Kennedy 1986a, p. 40, Pls. 3; 4:4–5; 5:1–3; Figs. 4A, 15–16 (holotype), 18). Microconchs of *P. (P.) oldhami* have yet to be recognized; macroconchs differ in the generally coarser ornament of *haldemsis*, with well-developed, rather than incipient bullae when young. Adults differ in the persistence of ornament in *haldemsis* in middle growth.

Occurrence. – Upper Upper Campanian, Köpinge, Sweden, Münster Basin and elsewhere in Germany; Gschliefgraben, Austria; Norfolk, England; Northern Ireland; Poland; Ukraine and Turkmenia.

Pachydiscus (Pachydiscus) cf. *subrobustus* Seunes, 1892

Fig. 12

Synonymy.. – compare: □1892 *Pachydiscus subrobustus* – Seunes, p. 15, Pl. 13 (4):1. □1894 *Pachydiscus subrobustus* Seunes – de Grossouvre, p. 200, Pl. 36:2. □1910 *Pachydiscus subrobustus* Seunes – Frech, p. 4, Pl. 1:1; Figs. 2–3. □1913 *Pachydiscus subrobustus* Seunes – Nowak, p. 357, Pl. 41:15. □1925 *Pachydiscus subrobustus* Seunes – Diener, p. 108. □1951 *Pachydiscus subrobustus* Seunes – Mikhailov, p. 70, Pl. 9:43–44. □1952 *Pachydiscus subrobustus* Seunes – Collignon, p. 92. □1955 *Pachydiscus subrobustus* Seunes – Collignon, p. 83. □1964 *Pachydiscus subrobustus* Seunes – Giers, p. 265, Pl. 5:3 (*pars*). □ *non* 1971 *Pachydiscus subrobustus* Seunes – Collignon, p. 34, Pl. 454:2411. □1974 *Pachydiscus subrobustus* Seunes – Naidin, p. 185, Pl. 65:1; Fig. 33. □1984 *Pachydiscus* cf. *subrobustus* Seunes – Kennedy & Summesberger, p. 161, Pl. 8:4. □1993 *Pachydiscus (Pachydiscus)* cf. *subrobustus* Seunes – Kennedy, p. 103, Pls. 1:1, 8–9; 2:13–14. □1993 *Pachydiscus (Pachydiscus) subrobustus* Seunes – Hancock & Kennedy, p. 161; ? Pl. 3:2–3.

Type. – Lectotype, by subsequent designation of Kennedy & Summesberger (1984, p. 161), is the original of Seunes (1892, Pl. 13 (4):1), from Tercis, Landes, France. The specimen has not been traced.

Description. – LO 7135t (Fig. 12) from Tosterup is a wholly septate composite mould, distorted into an ellipse, with a maximum preserved diameter of 119 mm. Coiling appears to have been moderately involute, the umbilicus deep, with a feebly convex subvertical wall, the umbilical shoulder broadly rounded. The umbilicus comprises 22% approximately of the diameter. Whorl section compressed oval, with greatest breadth just outside the umbilical shoulder; the whorl breadth to height ratio is 0.75, but degree of compression may have been modified by *post-mortem* compaction. On the earlier parts of the outer whorl ribs arise at the umbilical seam,

Fig. 12. Pachydiscus (Pachydiscus) cf. *subrobustus* Seunes, 1892. LO 7135t, from the uppermost Lower Campanian – lower Upper Campanian of Tosterup. All ×1.

sweep back across the umbilical wall, strengthen on the umbilical shoulder, and appear to develop into sharp umbilical bullae (preservation is defective) which give rise to pairs of ribs, with occasional intercalated ribs between; there are long intercalated ribs. On the last part of the outer whorl most ribs are primaries, with an estimated total of 46 ribs per whorl. The ribs are sharp, narrow, and separated by wide interspaces, straight and prorsiradiate on the inner and middle flank, then flexing forwards and concave on the outer flank and flexed back and broadly convex across the venter.

Occurrence. – The locality Tosterup ranges from uppermost Lower to lower Upper Campanian. The type material of *P. (P.) subrobustus* is from the Upper Campanian of Tercis, Landes, France, and there are also records from the Upper Campanian of the Gschliefgraben, Austria; Pontus, Turkey; Poland, Ukraine, and the Münster Basin, Germany.

Genus *Patagiosites* Spath, 1953

Type species. – *Ammonites patagiosus* Schlüter, 1867, p. 22, Pl. 4:4–5, by original designation of Spath (1953, p. 38); = *Ammonites stobaei* Nilsson, 1827, p. 5, Pl. 1:1–2.

Patagiosites stobaei (Nilsson, 1827)

Figs. 13–17

Synonymy. – □1827 *Ammonites stobaei* – Nilsson, p. 5, Pl. 1:1–2. □1867 *Ammonites patagiosus* – Schlüter, p. 22, Pl. 4:4–5. □?1872 *Ammonites stobaei* Nilsson – Schlüter, p. 56 (*pars*) Pl. 17:6–7; *non* Pl. 17:4–5 [=*P. (P.) lundgreni* de Grossouvre, 1894]; *non* Pl. 18:10–11 [=*P. (P.) pseudostobaei* (Moberg, 1885)]. □1872 *Ammonites patagiosus* – Schlüter, p. 66, Pl. 20:7–8. □1885 *Ammonites stobaei* Nilsson – Moberg, p. 18, Pl. 2:1–5 (with additional early synonymy). □?1889 *Ammonites (Pachydiscus) stobaei* Nilsson – Griepenkerl, p. 100. □1894 *Desmoceras stobaei* Nilsson,

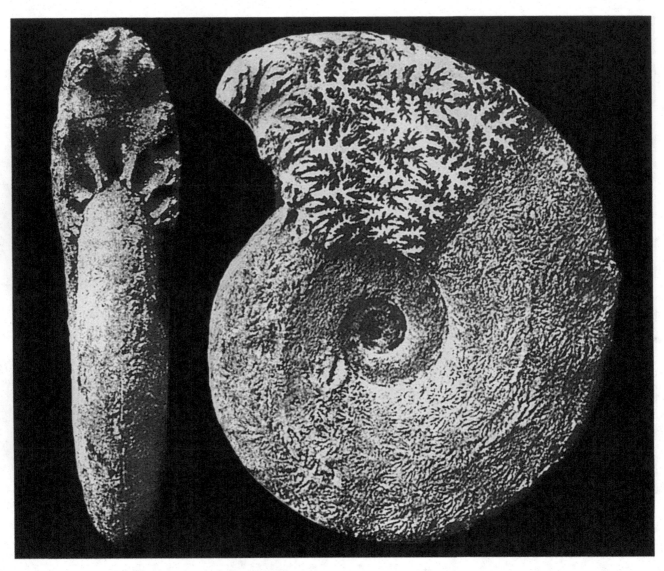

Fig. 13. Patagiosites stobaei (Nilsson, 1827). Copy of Moberg's original figure of the lectotype (1885, Pl. 2:1), LO 14T, probably middle Upper Campanian of Köpinge. ×0.25.

sp. – de Grossouvre, p. 170. □1907 *Pachydiscus patagiosus* Schlüter; Paulcke, p. 62. □1913 *Pachydiscus patagiosus* Schlüter – Nowak, p. 344. □1925 *Parapuzosia stobaei* Nilsson – Diener, p. 130. □1947 *Puzosia (Parapuzosia) stobaei* (Nilsson) – Hägg, p. 93. □1953 *Patagiosites patagiosus* (Schlüter) – Spath, p. 38. □1954 *Pachydiscus (Parapachydiscus) egertoni* (Forbes) – Hägg, p. 56, Pl. 9:98. □1954 *Puzosia (Parapuzosia) stobaei* (Nilsson) – Hägg, p. 56. □1964 *Pachydiscus stobaei* (Nilss.) – Giers, p. 258, Pl. 4:1–4; Fig. 4. □1964 *Pachydiscus koeneni* Grossouvre – Giers, p. 263 (*pars*), *non* Pl. 5:1 [=*P. (P.) haldemsis* (Schlüter)]; Pl. 5:2. □1964 *Pachydiscus patagiosus* (Schlüt.) – Giers, p. 267, Pl. 5:4; Fig. 6. □1974 *Pachydiscus stobaei* (Nilsson); Naidin, p. 184, Pls. 67:2; 68:2. □1983 *Parapuzosia stobaei* (Nilsson) – Regnell, p. 61; Fig. 4. □1988 *Pachydiscus stobaei sensu* Giers – Jagt, Pls. 1b; 2a–

b. □1993 *Parapuzosia (Parapuzosia) stobaei* (Nilsson) – Kennedy, p. 104.

Type. – Lectotype, here designated, is the original of Nilsson (1827, p. 5, Pl. 1:1–2), LO 14T, from the middle Upper Campanian of Köpinge, Sweden, refigured by Moberg (1885, Pl. 2:1), Regnell (1983, Fig. 4) and herein as Figs. 13–14.

Description. – Early growth stages, to a diameter of 67 mm approximately, are represented by the original of Moberg (1885, Pl. 2:5), from Köpinge, LO 720t (Fig. 15C–D). The specimen is a crushed, wholly septate composite internal mould. Coiling is fairly involute, with 70% of the previous whorl covered. The umbilicus comprises 27% of the diameter and is shallow, with a low flattened wall and narrowly rounded umbilical shoulder. The whorl section is

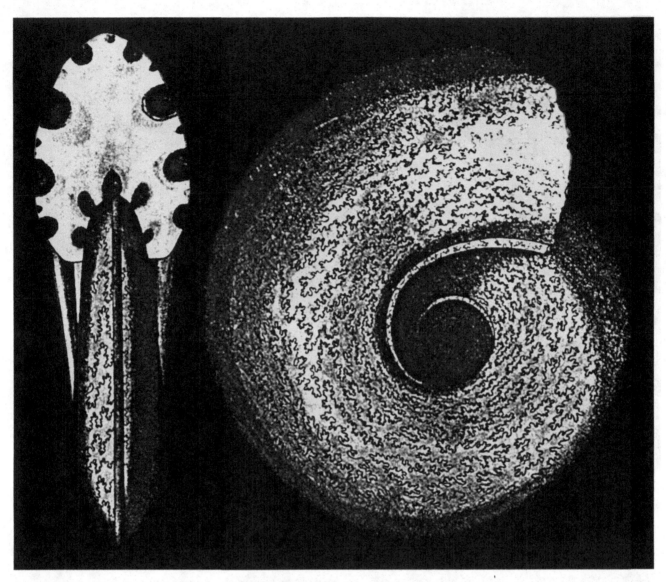

Fig. 14. Patagiosites stobaei (Nilsson, 1827). Copy of Nilsson's original figure of the lectotype (1827, Pl. 1:1–2), LO 14T, probably middle Upper Campanian of Köpinge. Approximately ×0.25.

compressed oval, with the greatest breadth just outside the umbilical shoulder and whorl breadth to height ratio 0.6, although this may have been modified by *post-mortem* crushing. The inner flanks are feebly convex, the outer flanks convergent, the venter narrowly rounded.

There are an estimated eight narrow straight prorsiradiate constrictions per whorl, deeply incised at the umbilical shoulder, and feebly convex across the venter. All are preceded by a narrow, markedly bullate collar rib, and succeeded by a subequal but non-bullate collar rib. The shell surface is smooth between constrictions, but for obscure traces of ribs, most obvious across the venter. Somewhat larger is the specimen figured by Hägg (1954, Pl. 9:98) from Tosterup, as *Pachydiscus (Parapachydiscus) egertoni* (Fig. 16E–F herein). The fragmentary nucleus of

this specimen is almost totally abraded, but for traces of a single constriction. The fragmentary outer whorl appears to be body chamber, and has a maximum preserved whorl height of 35 mm. The whorl breadth to height ratio is 0.8, the whorl section oval, with greatest breadth just below mid-flank. Two strong constrictions are present on the fragment, with strong adapical bullate constrictions and weak non-bullate adapertural constrictions. There are two distant coarse ribs on the adapical part of the fragment. Moberg (1885, Pl. 2:2; LO 719t) figured a very worn but undeformed phragmocone from Köpinge (Fig. 16A–D) with the following dimensions: D=105.5 (100); Wb=41.2 (39.1); Wh=55.0 (52.1); Wb:Wh=(0.75); U=26.4 (25.0). The inner whorls show five constrictions per half whorl; there are traces of two constrictions on the

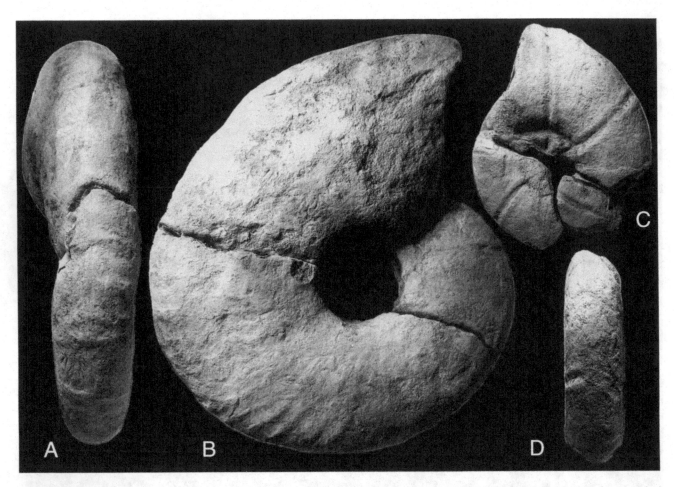

Fig. 15. Patagiosites stobaei (Nilsson, 1827), probably middle Upper Campanian, Köpinge. □A–B. LO 7136t, ×0.8. □C–D. LO 720t, the original of Moberg (1885, Pl. 2:5), ×1.

adapical part of the outer whorl, but no other traces of ornament survive. A further phragmocone, LO 7136t from Köpinge, lacks a nucleus and is distorted, but preserves ornament as a composite mould (Fig. 15A–B). There are three prorsiradiate, straight constrictions on the adapical 60° of the outer whorl, flanked by a bullate adapical and weaker, non-bullate adapertural collar-rib. There are obscure traces of ribs between the constrictions, while most of the outer whorl is ribbed. Ribs arise as mere striae on the inner flank and are markedly prorsiradiate. They strengthen markedly on the outermost flank, where they are feebly concave, sweeping forwards and strengthening across the ventrolateral shoulder and crossing the venter in a broad convexity. This ornament extends to the largest preserved diameter, 140 mm.

The lectotype has the following dimension: D=425 (100); WB=110 (25.9), Wh 172 (40.5); WB:WH=0.64; U=130 (30.6). It and other large specimens in the Lund Collections are smooth, with neither ribs nor constrictions.

Discussion. – We regard *Pachydiscus patagiosus, P. koeneni* and *P. stobaei* of Giers as synonyms of Nilsson's *Ammonites stobaei*, and also regard Schlüter's *Ammonites patagiosus* as a further synonym, following revision of their material from the Münster Basin (Kaplan, Kennedy & Ernst, unpublished).

Specimens referred to *P. patagiosus* by Giers are all composite moulds, mostly crushed. Early phragmocone whorls show coiling to have been fairly involute, with 70% of the previous whorl covered. The umbilicus is of moderate width and depth, with a broadly rounded umbilical wall and shoulder. The whorls expand slowly with, in the least-deformed specimens, a whorl breadth to height ratio of 0.96, the inner flanks very broadly rounded, the outer flanks flattened and convergent, the ventrolateral shoulders broadly rounded, the venter broad and feebly convex. There are typically four to five narrow, distant, straight, feebly prorsiradiate constrictions per half whorl. These are flanked by a narrow, weak, prominently bullate adapical, and non-bullate adaper-

Fig. 16. Patagiosites stobaei (Nilsson, 1827). □A–D. LO 719t, the original of Moberg (1885, Pl. 2:2), probably middle Upper Campanian of Köpinge. □E–F. RM unregistered, the original of *Pachydiscus (Parapachydiscus) egertoni* Forbes of Hägg (1954, Pl. 9:98), loc. CV5 of Tosterup, uppermost Lower – lower Upper Campanian. All ×1.

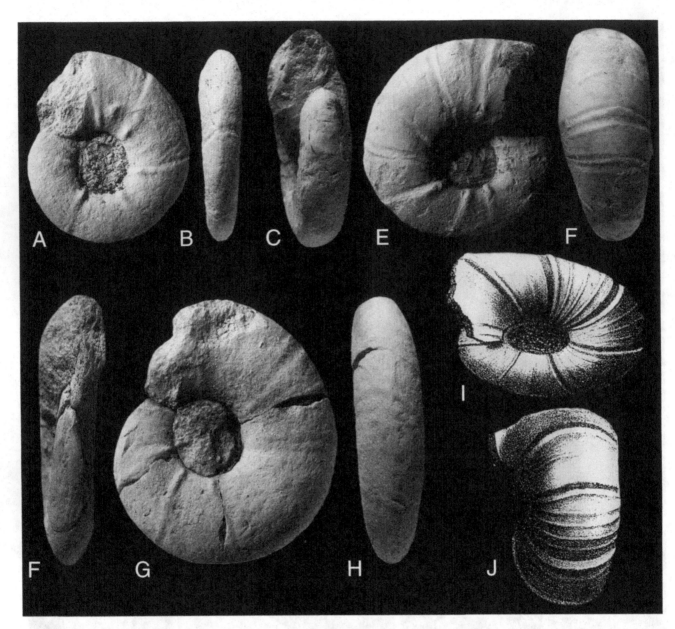

Fig. 17. Patagiosites stobaei (Nilsson, 1827). □A–B. Paralectotype of *Patagiosites patagiosus* (Schlüter, 1867), IPB 22c, the original of Schlüter (1867, Pl. 4:5), from the Upper Campanian of Coesfeld, Westphalia, Germany. □C–E. Lectotype of *Patagiosites patagiosus* (Schlüter, 1867), IPB 22b, the original of Schlüter (1867, Pl. 4:4), from the Upper Campanian of Coesfeld, Westphalia, Germany. □F–H. Paralectotype of *Patagiosites patagiosites* (Schlüter, 1867), IPB 22a, from the Upper Campanian of Sükerhook bei Coesfeld, Westphalia, Germany. □I–J. Copy of Schlüter (1872, Pl. 20:7–8), the original of which was from the Upper Campanian of Darup, Westphalia, Germany. All ×1.

tural collar-rib. The shell between the ribs is smooth in the early whorls of the smallest specimens, but as size increases, occasional low flank ribs develop. During the later growth stage of this group of specimens, up to three long or short ribs appear between constrictions. They are bullate or not, straight and prorsiradiate on the flanks, may weaken at mid-flank, then strengthen on the outer flank and cross the venter in a broad convexity. Collar-ribs strengthen, and constrictions become less prominent in the largest specimens of this group, where ornament is

of distant, coarse, bullate, straight prorsiradiate ribs flanked by feeble constrictions adapically, with or without associated adapertural ribs and up to four long and short ribs between them, the short ribs commonly extending across the ventrolateral shoulders and venter only. Such individuals are up to 100 mm in diameter, and some seem to be adult and thus possible microconchs, as with Hägg's *P. (P.) egertoni* (Fig. 16E–F).

Specimens identified as *Pachydiscus koeneni* by Giers have inner whorls like those of his *P. patagiosus*, with constrictions and associated collar-ribs, of which traces remain on the inner whorls of larger individuals, as with the Köpinge specimen shown as Fig. 15A–B. These are in general poorly preserved and very crushed. Coiling is involute, the umbilicus small, shallow (accentuated by crushing) with a broadly rounded wall and umbilical shoulder. The original whorl section appears to have been compressed, with flattened, subparallel flanks and broadly rounded ventrolateral shoulders and venter, which are crossed by ribs and constrictions in a broad convexity. There are 38 ribs per half whorl in the best-preserved specimen, and ornament of this type extends to around 200 mm diameter. The largest specimens are septate to around 350 mm diameter. Badly crushed, they bear distant coarse primary ribs, straight on the inner to middle flank, and feebly concave on the outer, with generally two short intercalated ribs between.

Specimens identified as *Pachydiscus stobaei* by Giers include large, wholly septate fragments with whorl heights of up to 180 mm, bearing distant bar-like straight, prorsiradiate ribs that arise at umbilical bullae, and weaken and efface on the ventrolateral shoulders and venter. The largest complete Münster Basin specimen is the original of Giers (1964, Pl. 4:1), with 14–15 ribs on the outer whorl (which is part body chamber) at a diameter of 800 mm.

The smallest of the Swedish specimens included in this species (Figs. 15C–D, 16E–F) differ in no significant respects from the surviving type material of *Patagiosites patagiosus* (Schlüter, 1867). The smallest of these, paralectotype IPB 22c, is the original of Schlüter (1867, Pl. 4:5), from Coesfeld, Westphalia (Fig. 17A–B). It is a crushed composite mould, with the following dimensions: D=58.9 (100); Wb=11.7 (19.8); Wh=21.1 (35.8); Wb:Wh=0.55; U=5.3 (9.0). Coiling is fairly involute, the umbilicus comprising 9% of the diameter. The original proportions cannot be reconstructed (because of crushing), but flanks and venter appear to have been broadly rounded. There are an estimated eight strong adapical collar ribs per whorl. These arise at the umbilical seam, are strong on the umbilical wall, and strengthen into sharp bullae on the umbilical shoulder. They give rise to a narrow, rounded rib, straight and prorsiradiate on the inner flank, weakening, flexed forward and feebly concave on the outer flank, and crossing the venter in a near-transverse course. They are succeeded by narrow constrictions on the last half of the outer whorl; constrictions are not obvious on the first half of the outer whorl. Constrictions may be succeeded by a much weaker adapertural collar rib, and there are traces of occasional non-bullate ribs between the constrictions. Paralectotype IPB 22a (Fig. 17F–H), from Sükerhook at Coesfeld, Westphalia is a paralectotype, again a crushed composite mould, with the following dimensions: D=67.1 (100); Wb=16.2 (24.1); Wh=25.5 (38.0); Wb:Wh=0.63; U=18.8 (28.0). To a

diameter of 53 mm it corresponds to the smaller paralectotype, with, on the last half whorl, prominent bullate adapical ribs associated with marked constrictions. Beyond 53 mm, the bullae become less conspicuous, ornament consisting of widely spaced feeble constrictions flanked by weak, equal collar ribs with weak bullae and traces of primary ribs in between that are conspicuous only on ventrolateral shoulders and venter. The lectotype is the original of Schlüter, 1867, Pl. 4:4; IPB 22b, shown here as Fig. 17C–E, from Coesfeld, Westphalia. It is preserved in hard grey limestone, as are Giers' specimens. It is a composite mould, distorted into an ellipse, with a maximum preserved diameter of 52 mm. Coiling is fairly involute, with 80% approximately of the previous whorl covered, the umbilicus comprising an estimated 31% of the diameter, quite deep, with a flattened wall. The original whorl section cannot be determined being in some places compressed, in others depressed, with feebly convex flanks, broadly rounded ventrolateral shoulders and a feebly convex venter. The first half of the outer whorl, to a diameter of 45 mm, corresponds to the previous specimen, with five constrictions and corresponding bullate adapical primary collar ribs. On the last half whorl there are also five constrictions, broader and more conspicuous than in the early growth stages, flanked by narrow, sharp ribs, with no, or weak bullae. The constrictions are in some cases doubled on the outer flanks, ventrolateral shoulders and venter, with traces of pairs of ribs and incipient constrictions between successive fully developed constrictions. This ribbing becomes progressively more conspicuous towards the adapertural end of the specimen. Ribs and constrictions are straight and prorsiradiate on the flanks and near-straight and transverse on the venter. *Patagiosites griffithi* (Sharpe, 1855) (p. 28, Pl. 11:3; Fig. 18 herein) is characterized by relatively high, ovoid whorls with a narrowly arched venter and only five to six deep constrictions per whorl, lacking conspicuous associated collar ribs and being smooth between constrictions at diameters where *P. stobaei* is ribbed. It is known from the Upper Campanian of Northern Ireland and Norwich, England. *Patagiosites amarus* (Paulcke, 1907) (p. 227, Pl. 20:5, 7) from Patagonia has depressed subcircular whorls, evolute coiling and strong constrictions with several intercalated ribs to 65 mm diameter, but thereafter lost, according to Paulke's figure; the *Patagiosites* aff. *amarus* of Spath (1953, p. 39, Pl. 10:7) from Graham Land, Antarctica, has similarly massive whorls with very coarse collar ribs and numerous weaker intercalatories to a larger size. *Patagiosites arbucklensis* (Anderson, 1958) (see revision in Matsumoto (1959, p. 60, Pls. 16:1; 17:1–2) is from the Campanian of California. It has numerous long ribs between the collar ribs at a very small size, whereas *P. stobaei* has few or none, is more evolute during later growth with subparallel flanks and very delicate flexuous flank ribbing. *Patagiosites alaskensis* Jones, 1963 (p. 45, Pls. 38–

Fig. 18. Patagiosites griffithi (Sharpe, 1855). Lectotype, GSM 37238, from the Upper Campanian White Limestone of Derry, Northern Ireland. All ×1.

40; 41:1, 3, 7, 9; Figs. 24–25) from the Maastrichtian of Alaska has markedly concave constrictions, five to seven per whorl on the early whorls, which are smooth or ribbed, the constrictions persisting to a larger size than in *P. stobaei*. The coiling is more evolute in middle growth, the whorls slower expanding, to give a quite different shell shape.

Occurrence. – The types are from the lower Upper Campanian of southern Sweden. The species has also been recorded from Donbass (Naidin 1974); Mons Basin, Belgium; Liège, Belgium, and southern Limburg, The Netherlands. In Germany the species ranges from uppermost Lower Campanian into the upper Upper Campanian.

Genus *Nowakites* Spath, 1922

Type species. – *Pachydiscus carezi* de Grossouvre, 1894, p. 190, Pls. 25:3; 37:5, by original designation of Spath (1922, p. 124).

Nowakites hernensis (Schlüter, 1867)

Fig. 19

Synonymy. – ☐1867 *Ammonites Hernensis* – Schlüter, p. 35, Pl. 6:4. ☐ *non* 1872 *Ammonites Hernensis* Schlüter – Schlüter, p. 40, Pl. 11:13–14 [=*Puzosia (Puzosia) muelleri*

Grossouvre]. ☐1885 *Ammonites* sp. – Moberg, p. 25, Pl. 2:6. ☐1922 *Tragodesmoceras hernensis* (Schlüter) – Spath, p. 128. ☐1925 *Tragodesmoceras hernense* (Schlüter) – Diener, p. 131. ☐ *non* 1953 *Tragodesmoceras hernense* Schlüt. – Ødum, p. 24, Pl. 4:3 [? = *Tragodesmoceras clypeale* (Schlüter, 1872)]. ☐1979 *Nowakites hernensis* (Schlüter) – Matsumoto, p. 37. ☐? 1991 *Nowakites* cf. *hernensis* (Schlüter) – Kennedy & Christensen, p. 210, Pl. 1:3. ☐1991 *Nowakites hernensis* (Schlüter) – Kennedy & Christensen, p. 210, Fig. 4.

Type. – Holotype, by monotypy, is IPB 27, from the 'Untersenonen grauen Mergeln des Schachtes von der Heydt bei Herne in Westphalen' (Fig. 19A–B).

Description. – There are a series of specimens from depths of 1328–1364 m in the Svedala borehole. The best-preserved of these, from a depth of 1364 m, is a crushed composite mould 62 mm in diameter (Fig. 19C). Coiling is moderately involute, with the apparently rather shallow umbilicus comprising 35% of the diameter. There are four prominent, very distant umbilical bullae on the 270° of the outer whorl preserved; these give rise to markedly concave prorsiradiate collar ribs that are succeeded by prominent, parallel constrictions. Between constrictions are numerous shorter, concave ribs, most obvious on the outer half of the flank, ventrolateral shoulders and venter.

The *Ammonites* sp. of Moberg (1885, p. 25, Pl. 2:6) is based on SGU Type 3874 (Fig. 19D–F). Hägg (1930) referred it to *Tragodesmoceras clypeale* (Schlüter, 1867),

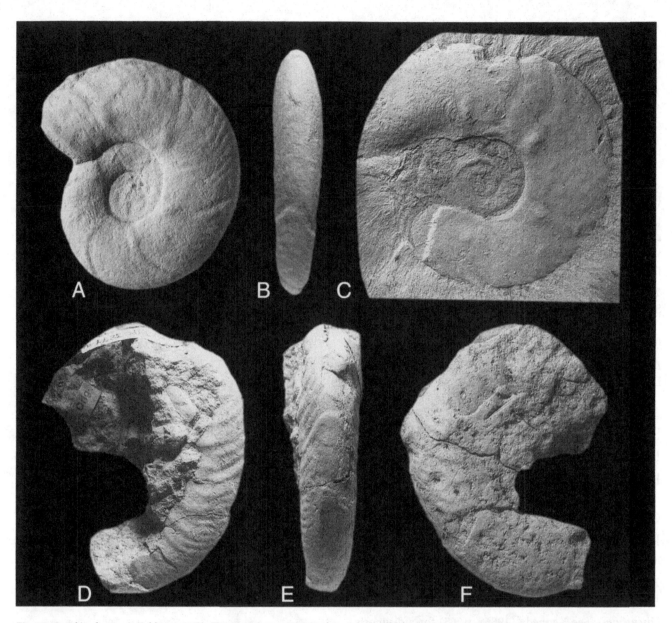

Fig. 19. Nowakites hernensis (Schlüter, 1867). □A–B. Holotype, IPB 27, the original of Schlüter (1867, p. 35, Pl. 6:4), from the 'Untersenonen grauen Mergeln des Schachtes von der Heydt bei Herne in Westphalen'. □C. SGU unregistered, from the Svedala borehole at a depth of 1364 m. □D–F. SGU Type 3874a, the original of *Ammonites* sp. of Moberg (1885, Pl. 2:6), from the upper Middle Santonian of Eriksdal. All ×1.

but the venter is broadly rounded with widely separated constrictions and associated adapical collar ribs. A pronounced ridge, displaced to one side of the mid-venter, is due to crushing. Only the outer flanks and venter are preserved, with well-developed concave ribbing.

Discussion. – The sparse constrictions, prominent bullae and very weak ribs between collar ribs readily separate *Nowakites hernensis* from the other member of the genus, as do its seemingly compressed, slowly expanding whorls, reminiscent of certain Kossmaticeratinae. *Nowakites lemarchandi* (de Grossouvre, 1894) (p. 173, Pl. 22:5) from the Santonian of the Corbières, France, has fewer con-

strictions, weaker bullae and much coarser ribs between the collar ribs. *Nowakites carezi* (de Grossouvre, 1894) (p. 190, Pl. 25:3) and *N. savini* (de Grossouvre, 1894) (p. 152, Pl. 25:4) are stouter shells with much coarser ribs, less markedly differentiated collar-ribs, and less prominent constrictions. *Nowakites paillettieanus* (d'Orbigny, 1841) (p. 339, Pl. 102:2) has near-even, coarser ribbing, without conspicuous bullae.

Occurrence. – Santonian of Herne, Westphalia and elsewhere in the Münster Basin, Germany, Eriksdal and the Svedala borehole, Sweden, and, possibly Bornholm, Denmark.

Family Muniericeratidae Wright, 1952

Genus *Tragodesmoceras* Spath, 1922

Type species. – *Desmoceras clypealoides* Leonhard, 1897, p. 57, Pl. 6:2, by original designation of Spath (1922, p. 127).

Tragodesmoceras clypeale (Schlüter, 1867)

Synonymy. – □1905 *Tragodesmoceras clypeale* (Schlüter) – Wegner, p. 207. □1922 *Tragodesmoceras clypeale* (Schlüter) – Spath, p. 128. □1925 *Tragodesmoceras clypeale* Schlüter – Diener, p. 131 (with additional synonymy). □?1953 *Tragodesmoceras hernense* Schlüt. – Ødum, p. 24, Pl. 4:3.

Types. – Lectotype, here designated, is the original of Schlüter (1872, Pl. 15:9–10, 13), from Salzberg bei Quedlinburg, as are the figured paralectotypes, which are in the collections of the Museum für Naturkunde, Berlin.

Discussion. – Ødum (1953) recorded this species from the Lower Santonian of the Höllviken-2 borehole.

Superfamily Hoplitaceae H. Douvillé, 1890
Family Placenticeratidae Meek, 1876

Genus and subgenus *Hoplitoplacenticeras* Paulcke, 1907

ICZN name no. 1345 (=*Dechenoceras* Kayser, 1924, p. 174)

Type species. – *Hoplites–Placenticeras plasticus* Paulcke, 1907, p. 186; ICZN Opinion 555, 1959: name no. 1629.

Discussion. – See Kennedy (1986a, p. 63) for a diagnosis and discussion of *Hoplitoplacenticeras* and its subgenera, plus a preliminary account of the German species of *H.* (*Hoplitoplacenticeras*).

Hoplitoplacenticeras (*Hoplitoplacenticeras*) cf. *coesfeldiense* (Schlüter, 1867)

Synonymy. – compare: □1867 *Ammonites Coesfeldiensis* – Schlüter, p. 14 (*pars*), Pl. 1:1, 4 only [*non* 2–3, ?=*H. vari*; *non* 5, ?=*H. dolbergense* (Schlüter, 1876)]. □1953 *Dechenoceras coesfeldiense* Schlüt. – Ødum, p. 23, Pl. 3:2. □1986a *Hoplitoplacenticeras coesfeldiense* (Schlüter) – Kennedy, p. 73, Pl. 9:9–10; Fig. 27B–C, F (with full synonymy).

Discussion. – Ødum (1953, p. 23, Pl. 3:2) figured a crushed fragment that may belong to this species from the Upper Campanian of the Höllviken-2 borehole at a depth of 611.12 m. We have not seen this specimen.

Occurrence. – *H.* (*H.*) *coesfeldiense* is known from the Upper Campanian of the Münster Basin in Germany, Ukraine, Russia, Turkmenia, Limburg, The Netherlands, and northern Aquitaine, France.

Suborder Ancyloceratina Wiedmann, 1966
Superfamily Turrilitaceae Gill, 1871
Family Nostoceratidae Hyatt, 1894

Genus *Bostrychoceras* Hyatt, 1900

Type species. – *Turrilites polyplocus* Roemer, 1841, p. 92, Pl. 14:1–2, by original designation of Hyatt (1900, p. 588).

Bostrychoceras polyplocum (Roemer, 1841)

Fig. 20

Synonymy. – □1841 *Turrilites polyplocus* – Roemer, p. 92, Pl. 14:1 only, *non* 2 [=*Eubostrychoceras saxonicum* (Schlüter, 1876)]. □ *non* 1953 *Bostrychoceras polyplocum* Roem. – Ødum, p. 22, Pl. 1:8 [?=*Nostoceras junior* (Moberg)]. □1986a *Nostoceras* (*Bostrychoceras*) *polyplocum* (Roemer); Kennedy, p. 92, Pl. 6:1; Pl. 15:1–3, 5–8; Figs. 32A–B; 33A–E; 34A–H; 35A–D (with full synonymy).

Fig. 20. *Bostrychoceras polyplocum* (Roemer, 1841). SGU unregistered, Upper Campanian, Trelleborg borehole, 580 m. ×1.

Type. – Lectotype, by subsequent designation of Kennedy (1986a, p. 95), is the original of Roemer (1841, Pl. 14:1) from an unspecified locality in northern Germany.

Description and discussion. – A fragment from the Trelleborg borehole at a depth of 580 m (Fig. 20) shows the crushed base and part of the flank of most of a whorl of specimen with an estimated whorl height of 22.5 mm. Ornament is of dense, narrow prorsiradiate ribs on the outer flank, linked at and intercalating between two rows of small tubercles (cf. Kennedy 1986a, Pl. 15:3–6; Fig. 33A, etc.). The specimen from the 'Lower' Campanian of the Höllviken-1 borehole at a depth of 804 m referred to this species by Ødum (1953, Pl. 1:8) is a small spire that has constrictions and seems to lack tubercles on the relatively few ribs, and may be closer to the Lower Campanian *Nostoceras junior* (Moberg, 1885) (see below).

Occurrence. – Upper Upper Campanian, Höllviken-1 borehole, Sweden, Northern Ireland, Norfolk, England, northern Aquitaine, France, Torallola, Montesqui and elsewhere in Catalonia; Navarra, Spain; Germany; Poland, Russia, Ukraine, Armenia, Kazakstan, Bulgaria, Iran, North Africa, Texas and northern Mexico.

Genus *Nostoceras* Hyatt, 1894

Type species. – *Nostoceras stantoni* Hyatt, 1894, p. 469, by original designation.

Discussion. – Genus *Mobergoceras* Schmid & Ernst, 1975, has as type species *Turrilites junior* Moberg, 1885 (p. 31, Pl. 3:14), by original designation. Schmid & Ernst (1975) regarded *Mobergoceras* as a subgenus of *Bostrychoceras* Hyatt, 1900. The illustrated material of *Mobergoceras* is limited to the holotype (Fig. 21), a specimen figured by Nowak (1913, Pl. 40:6) as *Heteroceras polyplocum*, a third figured by Schmid & Ernst (1975, Pl. 4:4), and a possible crushed individual figured by Ødum (1953, Pl. 1:7). This material differs from *Bostrychoceras polyplocum* and allied species in its small size, coarse ribbing, prominent flared constrictions, coiling with whorls in tight contact, apparent lack of an extended uncoiled early growth stage and of a loosely coiled, recurved adult body chamber. In these respects it resembles certain Campanian *Nostoceras* Hyatt, 1894, such as *N. saundersorum* Stephenson, 1941 (p. 46, Pl. 83:6–8). Accordingly, *Mobergoceras* is here regarded as a synonym of *Nostoceras* rather than a subgenus of *Bostrychoceras*.

Fig. 21. Nostoceras junior (Moberg, 1885). Holotype, LO 732T, the original of Moberg (1885, Pl. 3:14), probably middle Upper Campanian of Köpinge. ×1.

Nostoceras junior (Moberg, 1885)

Fig. 21

Synonymy. – □1885 *Turrilites junior* – Moberg, p. 31, Pl. 3:14. □?1913 *Heteroceras polyplocum* Roemer – Nowak, p. 385, Pl. 40:6; Pl. 45:48. □? 1953 *Bostrychoceras polyplocum* Roem. – Ødum, p. 22, Pl. 1:8. □1965 *Bostrychoceras (Mobergoceras) junior* (Moberg) – Schmid & Ernst, p. 342, Pl. 4:4.

Type. – Holotype, by monotypy, is the original of Moberg (1885, Pl. 3:4), LO 732T, from Köpinge, Sweden (Fig. 21 herein).

Description. – The holotype consists of 1.3 whorls of body chamber with the adult aperture preserved, the whorls in contact throughout, and the maximum preserved whorl height 14.5 mm. The upper whorl face is concave, but too poorly preserved to reveal details of ornament. The outer whorl face is convex, with 16–18 coarse, narrow prorsiradiate straight to feebly sinuous ribs that extend across the lower whorl face, where they are straight and radial, declining in strength towards the umbilicus. There are three, perhaps four constrictions on the fragment, each preceded by a markedly strengthened rib. The aperture is prolonged into a rostrum and appears somewhat contracted.

Discussion. – The coarse, distant ribbing and constrictions readily distinguish *N. junior* from other, nontuberculate *Nostoceras* such as *N. saundersorum* Stephenson, 1941 (p. 46, Pl. 83:6–8), which has flatter whorls, a shallow interwhorl suture, and numerous delicate ribs. *Nostoceras platycostatum* Kennedy & Cobban, 1993a (p. 131, Pl. 2:16–17; Pl. 4:1–13, 33, 34, 37A) has a much wider apical angle, with numerous near-transverse ribs, and only ill-defined constrictions. Most other *Nostoceras* species are tuberculate.

Occurrence. – Upper Campanian of Köpinge, Sweden; 'Lower' Campanian of the Höllviken-1 borehole, Sweden; uppermost Lower Campanian, Misburg, Germany; Campanian of Kaliszanay, Poland.

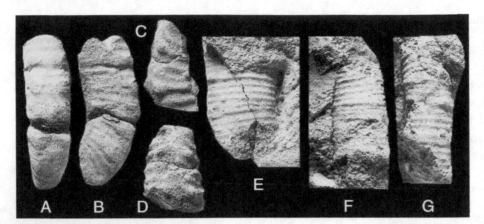

Fig. 22. Lewyites elegans (Moberg, 1885). □A–B. Holotype, LO 731T, the original of Moberg (1885, Pl. 3:10), probably middle Upper Campanian of Köpinge. □C–D. LO 733t, the original of *Helicoceras?* sp. of Moberg (1885, p. 33), probably middle Upper Campanian of Köpinge, Sweden. □E–G. SGU unregistered, uppermost Lower – lower Upper Campanian of Tosterup. All ×1.

Family Diplomoceratidae Spath, 1926
Subfamily Diplomoceratinae Spath, 1926

Genus *Lewyites* Matsumoto & Miyauchi, 1984

Type species. – *Idiohamites*(?) *oronensis* Lewy, 1969, p. 127, Pl. 3:10–11, by original designation by Matsumoto & Miyauchi (1984, p. 64).

Lewyites elegans (Moberg, 1885)
Figs. 22–23

Synonymy. – □1885 *Scaphites?* – Moberg, p. 30, Pl. 3:11. □1885 *Ancyloceras? elegans* – Moberg, p. 30, Pl. 3:10. □1885 *Helicoceras?* sp. Moberg, p. 33. □1913 (?)*Anisoceras elegans* – Moberg sp. – Nowak, p. 384, Pl. 40:7. □1986a *Neocrioceras* (*Schlueterella*)? *elegans* (Moberg) – Kennedy, p. 102, Pl. 17:3–6.

Type. – Holotype, by monotypy, is the original of Moberg (1885, Pl. 3:10), LO 731T (Fig. 22A–B herein), probably from the middle Upper Campanian of Köpinge, Sweden.

Description. – The holotype is a curved sector with part of a straight shaft of a phragmocone 41.5 mm long, with a maximum preserved whorl height of 16 mm. The whorl section is compressed oval in intercostal section, the venter more narrowly rounded than the dorsum, with a whorl breadth to height ratio of 0.84. The dorsum is very worn but appears to have been ornamented by delicate feebly convex ribs. These strengthen markedly across the dorsolateral margin, and are narrow and markedly rursiradiate on the flank of the curved sector, less so on the shaft, where the rib index is eight. Ribs join in pairs at coarse ventral tubercles that are linked across the venter by a pair of delicate looped ribs. One, two, and perhaps three nontuberculate ribs separate tuberculate ribs and

are narrow and transverse across the venter. A second fragment from Tosterup is in the SGU Collections (Fig. 22E–G). This has a whorl breadth to height ratio of 0.7 approximately with a maximum preserved whorl height of 17 mm. The rib index is 12, with delicate ventral tubercles on some ribs, and up to three nontuberculate ribs between. A somewhat larger fragment of venter is the original of Moberg's *Helicoceras* sp. from Köpinge (1885, p. 33; LO 733t; Fig. 22C–D herein); well-preserved, it has sharp tubercles, without clear ribs linking them across the venter, and one or two nontuberculate ribs separating the tuberculate ones. The largest fragment, also from Köpinge, is the original of Moberg's *Scaphites* (1885, p. 30; Pl. 3:11; LO 730t; Fig. 23 herein). This consists of part of the flank, with one row of ventral tubercles present. The maximum preserved whorl height is 34.5 mm, the rib index 11, with pairs of ribs linked at coarse tubercles, with one to three nontuberculate ribs between tuberculate groups.

Discussion. – There is some variation in the rib index of the present material, which is treated as belonging to a single species. *Lewyites oronensis* (Lewy, 1969) (p. 127, Pl. 3:10–11) from the Upper Campanian Mishash Formation of Israel is a close ally, with ribs joined in pairs at ventral tubercles but only a single nontuberculate rib between. With further and better material from Sweden it may well prove to be junior synonym of *elegans*. *Lewyites circularis* (Lewy, 1969) (p. 128, Pl. 3:9; Pl. 4:3 only), also from the Upper Campanian Mishash Formation of Israel, has a circular whorl section, similar flank ornament, but ribs that weaken or disappear on the venter. *Lewyites clinensis* (Adkins, 1929) (p. 208, Pl. 6; 10–11) from the Upper Campanian Anacacho Limestone near Cline, Uvalde County, Texas, has a nearly circular intercostal section, rib index of seven, flank ribs linked in pairs at ventral tubercles, and tubercles linked across the venter by pairs of weakened ribs, on the adapical of the holotype these support a low siphonal node. A single nontuberculate rib

Fig. 23. Lewyites elegans (Moberg, 1885). LO 730t, the original of *Scaphites?* of Moberg (1885, Pl. 3:11), probably middle Upper Campanian of Köpinge. All ×1.

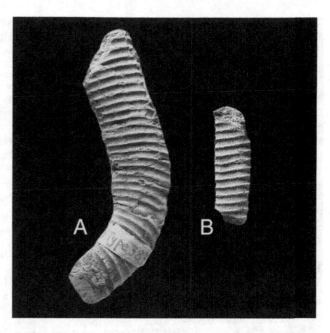

Fig. 24. Glyptoxoceras crispatum (Moberg, 1885). □A. Lectotype, original of Moberg (1885, Pl. 3:12), SGU Type 3877. □B. Paralectotype, SGU Type 3876. Both specimens are from the upper Middle Santonian of Eriksdal. All ×1.

separates tuberculate ribs. *Lewyites taylorensis* (Adkins, 1929) (p. 209, Pl. 6:12–13), from the Upper Campanian Pecan Gap Chalk of northeast Texas, is known from poor phosphatic fragments only with pairs of strong ribs linked at ventral tubercles with two weaker, annular, nontuberculate ribs between. It remains inadequately known.

Occurrence. – Upper Campanian of Köpinge and Tosterup, Sweden, northern Aquitaine, France, and Waganowice, Poland.

Genus *Glyptoxoceras* Spath, 1925

Type species. – *Hamites rugatus* Forbes, 1846, p. 117, by original designation.

Glyptoxoceras crispatum (Moberg, 1885)
Fig. 24A–B

Synonymy. – □1885 *Anisoceras* (*Hamites?*) *crispatus* – Moberg, p. 32, Pl. 3:12–13.

Types. – Lectotype, here designated, is the original of Moberg (1885, Pl. 3:12), from the Santonian of Eriksdal, Sweden, SGU Type 3877 (Fig. 24A). Paralectotype is SGU Type 3876 (Fig. 24B).

Description. – The lectotype is a crushed, partially septate composite mould 77 mm long and gently curved, with a maximum preserved whorl height of 18 mm. The rib index is eight, the ribs narrow, sharp, feebly convex on the phragmocone, less so on the adapical body chamber, and near-transverse, straight, and feebly rursiradiate at the adapertural end.

Discussion. – Kennedy (1986a) discussed this species in the context of *Glyptoxoceras retrorsum* (Schlüter, 1872) (p. 97, Pl. 30:5–10) of the Upper Campanian, noting, however, that it was much older. Small specimens of *retrorsum* (Schlüter, 1872, Pl. 30:8; Kennedy 1986a, Fig. 38D; Kennedy 1993, p. 108, Pl. 4:1–9, 11–19, 25, 26; Fig. 4) all have ribs that are feebly concave on the outer flank, while larger specimens suggest an even, elliptical or circular coil. *Glyptoxoceras aquisgranense* (Schlüter, 1872) (p. 102, Pl. 31:6–9; Kennedy *et al.* 1992, p. 274, Pls. 1:6–7, 11–12, 14–19; 2:1–5, 9–15; 3:1–9) of the Lower Campanian has a complex coiling ontogeny, although at the size represented by the lectotype of *G. crispatum* it has a circular to elliptical coil with a rib index of up to nine, but decreasing to five or six on some body chambers, the ribs rursiradiate, straight on the dorsal flank but flexed forwards in many specimens and feebly concave on the ventral part of the flank. *Glyptoxoceras souqueti* Collignon, 1983 (p. 186, Pl. 1:4) from the Upper Santonian of the Corbières has a variable rib index, from seven in the holotype to up to ten in other, as yet undescribed specimens, and is a close ally, the ribs being straight to feebly convex and prorsi- to rursiradiate (Fig. 25).

Occurrence. – As for type.

Fig. 25. Holotype of *Glyptoxoceras souqueti* Collignon, 1983, from the Upper Santonian, 'chemin de Sougraine aux Croutets', Corbières, Aude, France. M. Bilotte Collection, Toulouse. ×1.

Family Baculitidae Gill, 1871

Genus *Baculites* Lamarck, 1799

Type species. – *Baculites vertebralis* Lamarck, 1801, p. 103, through subsequent designation by Meek (1876, p. 391).

Baculites suecicus Moberg, 1885

Fig. 26A

Synonymy. – □1885 *Baculites suecicus* – Moberg, p. 34, Pl. 4:1.

Type. – Holotype, by monotypy, is LO 734T, the original of Moberg, 1885, Pl. 4:1 (Fig. 26A herein). Moberg noted that the specimen was labelled Köpinge, an Upper Campanian locality, but suggested on the basis of the matrix that it may have come from Kåseberga, a Lower Campanian locality.

Description. – The holotype is a very crushed composite mould of a phragmocone with a maximum preserved whorl height of 53 mm. The original whorl proportions cannot now be determined. There is an obscure ornament of growth lines, striae and low undulations, but no clearly differentiated ribs. The ornament defines a transient aperture with short dorsal and long ventral rostrum, and markedly concave at mid-flank. Only traces of the suture line survive, with quite deeply incised bifid lobes and saddles.

Discussion. – The holotype has so few diagnostic features that it cannot be usefully compared with other feebly ornamented *Baculites* species. In the absence of knowledge of intraspecific variation and original, undeformed whorl section, *suecicus* is best treated as a *nomen dubium*.

Baculites incurvatus Moberg, 1885, *non* Dujardin, 1837

Fig. 26C–E

Synonymy. – □1885 *Baculites incurvatus* Dujardin – Moberg, p. 36, Pl. 4:2–4.

Discussion. – *Baculites incurvatus* Dujardin, 1837, is a Coniacian species, revised by Kennedy (1984, p. 143, Pls. 32:12, 15–19; 33:1–22; Figs. 41, 42F–M). Moberg (1885) referred a series of specifically indeterminate fragments to the species. His Pl. 4:2, SGU Type 3878, is from the upper Middle Santonian of Eriksdal (Fig. 26C). It is 37.5 mm long with a maximum preserved whorl height of 13 mm, it has two prominent dorsolateral nodes preserved, and prominent markedly prorsiradiate lateroventral growth lines striae and riblets. The original of his Pl. 4:3, LO 735T, shown here as Figure 26D, is from the lower Lower Campanian of Kåseberga. It is 34 mm long, with a maximum preserved whorl height of 9 mm. Ornament is of two widely separated dorsolateral nodes which give rise to single concave crescentic ridges, projected markedly on the ventrolateral margin. The specimen may belong to the widely occurring group of *Baculites capensis* Woods, 1906

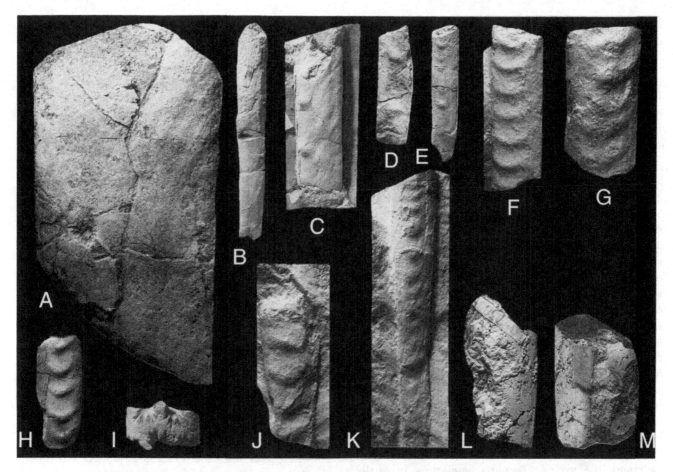

Fig. 26. □A. *Baculites suecicus* Moberg, 1885. Holotype, LO 734T, from the lower Lower Campanian, Kåseberga?, the original of Moberg (1885, Pl. 4:1). □B. *Baculites* sp., SGU Type 3881, the original of Moberg (1885, Pl. 4:7), from the upper Middle Santonian of Eriksdal. □C. *Baculites incurvatus* Moberg, 1885 *non* Dujardin, 1837. SGU Type 3878, the original of Moberg (1885, Pl. 4:2), from Eriksdal. □D. *Baculites incurvatus* Moberg, 1885 *non* Dujardin, 1837. LO 735t, the original of Moberg (1885, Pl. 4:3), from the lower Lower Campanian, Kåseberga. □E. *Baculites incurvatus* Moberg, 1885 *non* Dujardin, 1837. LO 736t, the original of Moberg (1885, Pl. 4:4). □F. *Baculites* cf. *aquilaensis* Reeside, 1927. MGUH 24152, Tosterup conglomerate, loc. CV2, Röd-mölla, uppermost Lower Campanian, □G *Baculites* cf. *aquilaensis* Reeside, 1927. MGUH 24153, Tosterup conglomerate, loc. CV2, Rödmölla, uppermost Lower Campanian. □H. *Baculites* cf. *aquilaensis* Reeside, 1927. RM Mo 135639, the original of Moberg (1885, Pl. 12:3), probably middle Upper Campanian of Köpinge. □I. *Baculites vertebralis* Moberg, 1885 *non* Lamarck, 1801. LO 737t, the original of Moberg (1885, Pl. 13:9), Åhus. □J. *Baculites* cf. *aquilaensis* Reeside, 1927. MGUH 24154. Tosterup conglomerate, loc. CV2, Rödmölla, uppermost Lower Campanian. □K. *Baculites* cf. *aquilaensis* Reeside, 1927. MGUH 24155. Tosterup conglomerate, loc. CV2, Rödmölla, upermost Lower Campanian. □L–M. *Baculites vertebralis* Moberg, 1885 *non* Lamarck, 1801. LO 738t, the original of Moberg (1885, Pl. 4:9), Balsberg, probably lower Upper Campanian. All ×1.

(p. 342, Pl. 44:6–7). The original of Moberg's Pl. 4:4, shown here as Fig. 26E, is from Eriksdal. It is 38 mm long, with a maximum preserved whorl height of 7.5 mm, the whorl section markedly trigonal as a result of *post-mortem* crushing. There are six delicate crescentic dorsolateral nodes and associated ribs. The fragment is again a specifically indeterminate member of the *capensis* group.

Baculites brevicosta? Moberg, 1885 *non* Schlüter, 1876

Synonymy. – □1885 *Baculites brevicosta* Schlüter? – Moberg, p. 37, Pl. 4:5–6.

Discussion. – Moberg (1885) compared two fragments from Eriksdal, a Santonian locality, to Schlüter's *brevicosta*, a Coniacian species. They fall within the variation range of the South African *Baculites capensis* group.

Baculites sp. Moberg, 1885

Fig. 26B

Synonymy. – □1885 *Baculites* sp. – Moberg, p. 37, Pl. 4:7.

Discussion. – This badly crushed fragment, SGU Type 3881, from the upper Middle Santonian of Eriksdal, is specifically indeterminate.

Baculites cf. aquilaensis Reeside, 1927

Fig. 26F–G, H, J–K

Synonymy. – compare: □1885 *Baculites anceps auctorum* (Lamarck, d'Orbigny, Schlüter, cet.) – Moberg, p. 37, Pl. 4:11–?12. □1927 *Baculites aquilaensis* – Reeside, p. 12, Pls. 6:11–13; 8:1–4; var. *separatus*, p. 12, Pls. 8:15–21; 9:6–15; 45:5–6; ?var. *obesus*, p. 12, Pl. 10:1–8. □1993 *Baculites aquilaensis* Reeside – Kennedy, p. 110, Pl. 4:22, 24.

Description and discussion. – We have only seen one of Moberg's specimens of *Baculites anceps* auctorum (Fig. 26H), but they certainly do not belong to the Upper Maastrichtian *Baculites anceps* Lamarck, 1822 (see revision in Kennedy 1986b, p. 58, Pls. 11:12–14; 12:7–11; Figs. 3E–H, 7A–C). Rather, they seem to belong to the same group as a series of poorly preserved fragments from the uppermost Lower Campanian Tosterup conglomerate shown in Figures 26F–G, J–K. These have a compressed ovoid whorl section when undeformed, the venter more narrowly rounded than the dorsum, the expansion rate low. Ornament is of feebly concave strong ribs, two or three in a distance equal to the whorl height, and extending across the dorsal two-thirds of the flanks. They are strongly projected forwards on the ventral third, weaken markedly and break down into riblets and striae, which may also intercalate between. Ornament effaces and crosses the venter in a narrow convexity.

Despite poor preservation, these specimens compare well with the variable *Baculites aquilaensis*, originally described from the Lower Campanian of the United States Western Interior, but also recorded from the Mons Basin, Belgium.

Occurrence. – Tosterup conglomerate, Köpinge, uppermost Lower – lower Upper Campanian.

Baculites vertebralis Moberg, 1885, *non* Lamarck, 1801

Fig. 26I, L–M

Synonymy. – □1885 *Baculites vertebralis* Lamarck – Moberg, p. 38, Pl. 4:8–9.

Discussion. – Moberg illustrated two specimens as *Baculites vertebralis* Lamarck, an Upper Maastrichtian species revised by Kennedy (1986b) and Birkelund (1993); neither of his specimens belong to the species. The original of his Pl. 4:8, LO 737t, from Åhus (Fig. 26I herein), is a single worn chamber with a maximum preserved whorl height of 20.8 mm. Better preserved is the original of his Pl. 4:9, LO 738T; Figs. 26L–M herein. It shows the adapical part of the specimen, a phosphatized fragment from Balsberg, preservation suggesting a lower Upper Campanian horizon. It is part-septate, 62 mm long, with a maximum preserved whorl height of 18 mm approximately, the whorl

section compressed oval. No ornament is preserved. The two specimens are specifically indeterminate.

Baculites angustus Moberg, 1885

Fig. 27A–E

Synonymy. – □1885 *Baculites angustus* – Moberg, p. 39, Pl. 4:10.

Type. – Holotype, by monotypy, is the original of Moberg (1885, Pl. 4:10), LO 739T, from Köpinge.

Description. – The holotype is a dark brown phosphatic internal mould of parts of four chambers. The fragment is 21.8 mm long, with a maximum preserved whorl height of 15.3 mm and whorl breadth to height ratio 0.67. The whorl section is ovoid, the venter more narrowly rounded than the feebly convex dorsum (Fig. 27D). Ornament is of delicate growth lines and riblets, near effaced on the dorsum, feebly concave across the dorsal half of the flanks and projected strongly forwards and effacing on the ventral half (Fig. 27B). The venter is worn. Suture line moderately incised, with broad-stemmed elements (Fig. 27E).

Discussion. – The locality Köpinge ranges from upper Lower to middle Upper Campanian, while this phosphatic fragment may be older. In the absence of better material it cannot be determined if *angustus* is a feebly ribbed species or a variant of a nodose species. It is best treated as a *nomen dubium*, being based on inadequate material.

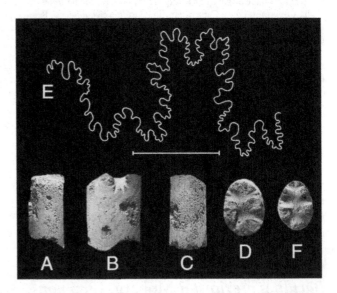

Fig. 27. □A–E. *Baculites angustus* Moberg, 1885. LO 739T, original of Moberg (1885, Pl. 4:10), a phosphatic fragment from Köpinge, probably middle Upper Campanian. □F. *Baculites schlueteri* Moberg, 1885. LO 741T, original of Moberg (1885, Pl. 4:14), a phosphatic fragment from Köpinge, probably middle Upper Campanian. Figs. A–D and F are ×1. Bar scale of suture line of *B. angustus* is 10 mm.

Fig. 28. Baculites anceps Lamarck, 1822. LO 740t, the original of *Baculites schlüteri*? of Moberg (1885, Pl. 4:13), from the Upper Maastrichtian Calcaire à *Baculites* of Picouville, Manche, France. All ×1.

Baculites schlueteri Moberg, 1885

Fig. 27F

Synonymy. – ☐1885 *Baculites schlüteri* – Moberg, p. 40, Pl. 4:14 only.

Type. – Lectotype, here designated, is the original of Moberg (1885, p. 40, Pl. 4:14), LO 741T, a phosphatized fragment from Köpinge.

Description. – The lectotype is a 14.5 mm long fragment with a maximum preserved whorl height of 12.4 mm and whorl breadth to height ratio of 0.74; parts of three chambers are preserved. The whorl section is markedly ovoid, with venter more narrowly rounded than dorsum (Fig. 27F). There are suggestions of delicate, concave flank ornament, but preservation is poor. Suture with moderately incised elements.

Discussion. – The fragment lacks the diagnostic features necessary to characterize the species, and in the absence of more adequate material should be regarded as a *nomen dubium*.

The specimen from Picouville, Manche, France, figured as a questionable *Baculites schlueteri* by Moberg (1885, Pl. 4:13), LO 740T, is shown as Fig. 28. It is a typical *Baculites anceps* Lamarck, 1822, from the Upper Maastrichtian (see revision in Kennedy 1986b, p. 58, Pls. 11:12–14; 12:7–11; Figs. 3E–H, 7A–C).

Baculites knorrianus (Desmarest, 1817)

Fig. 29

Synonymy. – ☐1817 *Baculites knorrianus* – Desmarest, p. 48, Pl. 1:3. ☐1987 *Baculites knorrianus* Desmarest – Kennedy & Summesberger, p. 32, Pls. 4:4–6; 5:1–15; Fig. 2 (with full synonymy). ☐1993 *Baculites knorrianus* Desmarest – Kennedy, p. 109, Pls. 5:13–22; 6:11–13; 18–23; Fig. 5A–C. ☐1993 *Baculites knorrianus* Desmarest – Birkelund, p. 52, Pl. 13:12–14 (with additional synonymy).

Type. – Neotype, by subsequent designation of Kennedy & Summesberger (1987, p. 33), is No. 7459a in the collections of the Naturhistorisches Museum Vienna, from the Lower Maastrichtian of Nagoryany, near Lvov in the Ukraine (Kennedy & Summesberger, 1987, Pl. 5:5, 7–8).

Description. – MGUH 24156, from Bjärnum, is a composite mould of phragmocone 82 mm long, with parts of six camerae preserved, the maximum whorl height 44.5 mm, the whorl breadth to height ratio 0.6, the whorl section compressed ovoid with venter more narrowly rounded than dorsum. The surface of the mould is smooth but for traces of markedly prorsiradiate riblets and grooves on the ventral part of the flanks. Suture line complexly incised with broad lobes and saddles (Fig. 29C).

Discussion. – Large size, compression, feeble ornament and sutural complexity show this fragment to be *Baculites knorrianus*, as revised by Kennedy & Summesberger (1987), Kennedy (1993) and Birkelund (1993).

Fig. 29. □A–C. Cast of *Baculites knorrianus* Desmarest, 1817. MGUH 24156, from the basal Lower Maastrichtian, Bjärnum. A–B are natural size. Bar scale of suture line is 10 mm.

Occurrence. – Lower Maastrichtian, Sweden, Denmark, Poland, Czechoslovakia, Ukraine; lower Upper Maastrichtian of Rørdal, Denmark (Birkelund 1993).

Baculites spp.

Fig. 30

Discussion. – Two further assemblages of *Baculites* are represented by a small number of fragments, both unfortunately lacking sufficient criteria for specific determination. The older assemblage is from the uppermost Lower Campanian of Ivö Klack (Fig. 30A–H). This includes fragments with whorl heights of up to 40 mm. Most are smooth, with a compressed ovoid whorl section; a few have growth lines, shallow furrows and riblets that define a transient aperture with short dorsal, and long ventral rostrum (Fig. 30E). A single fragment has crescentic dor-

solateral ribs and may be a separate species or a variant (Fig. 30D). Most of these specimens resemble the smooth *Baculites* known from the Campanian of the U.S. Western Interior (e.g., Cobban 1962, p. 714, Pl. 108:1–4; Fig. 1I–J).

A second, younger assemblage, probably from the middle Upper Campanian of Köpinge (Fig. 30I–L), has individuals with whorl heights of up to 39 mm, one with the aperture seemingly preserved (Fig. 30L). The dorsum is broad and flattened and the venter narrowly arched when compared with the Ivö assemblage, and ornament weak to absent.

Aptychi of *Baculites*

Fig. 31

Discussion. – The middle Upper Campanian at Köpinge has yielded a series of complete and fragmentary ammo-

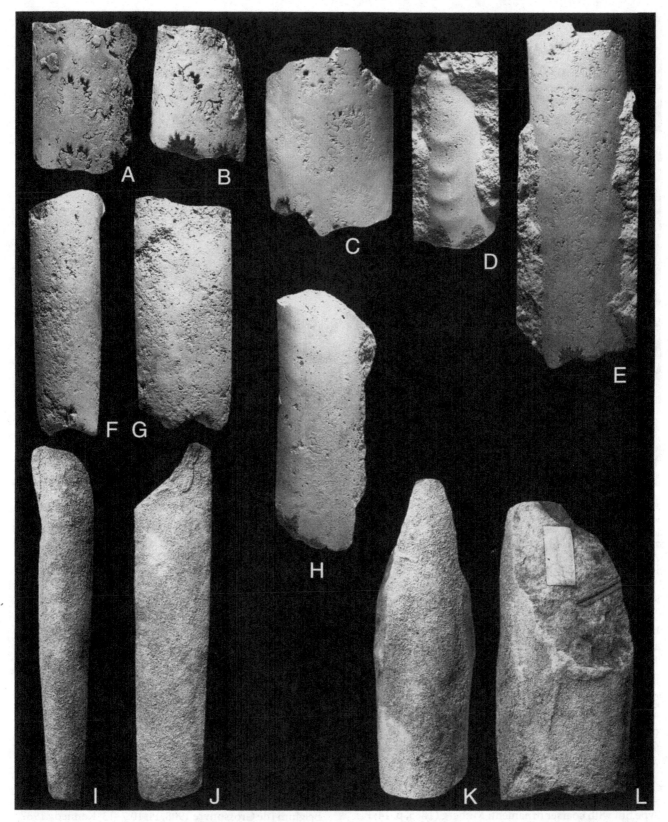

Fig. 30. ☐A–H. *Baculites* sp. from the uppermost Lower Campanian of Ivö Klack. ☐A–B. MGUH 24157; ☐C. SGU Collections. ☐D. SGU unregistered. ☐E. SGU unregistered. ☐F–G. MGUH 24158. ☐H. SGU unregistered. ☐I–L. *Baculites* sp. from Köpinge, probably middle Upper Campanian. ☐I–J. SGU unregistered. ☐K–L. SGU unregistered. All ×1.

Fig. 31. Rugaptychus. □A. Syntype of *Rugaptychus rugosus* (Sharpe, 1857, Pl. 26:9), Norwich Castle Museum unregistered, from the Upper Chalk of Norwich, England. □B. *Rugaptychus flexus* (Moberg, 1885), LO 748T, the original of Moberg (1885, Pl. 1:19), from the uppermost Lower Campanian or lower Upper Campanian of Balsberg. □C–H. *Rugaptychus rugosus* Sharpe of Moberg (1885), Köpinge, probably middle Upper Campanian. □C. LO 744t, original of Moberg (1885, Pl. 1:16). □D. LO 746t, original of Moberg (1885, Pl. 11:18). □E. LO 743t, original of Moberg (1885, Pl. 1:15). □F. LO 747t, original of Moberg (1885, Pl. 6:26). □G. LO 745t, original of Moberg (1885, Pl. 1:17). □H. LO 742t, original of Moberg (1885, Pl. 1:14). A, E and H, approximately ×1; B–D, F–G ×2.

nite jaws, referable to the form genus *Rugaptychus* Trauth, 1927. Schlüter (1876, Pl. 39:16) figured a specimen of *Baculites* from the Lower Maastrichtian Mucronatenkreide of Lüneburg, Germany with *Rugaptychus* in place, demonstrating the association. The type species of *Rugaptychus* by the subsequent designation of Moore (1957, p. 468) is *Aptychus rugosus* Sharpe (1857, p. 57, Pl. 24:8–9), from the Upper Campanian of Norwich, England, the more complete syntype of which is shown as Fig. 31A. As Sharpe (1857, p. 58) noted, Hébert (1856, Pl. 28:6) described, as *Aptychus insignis*, an aptychus of the same type but with coarser ornament. Moberg (1885, p. 14; Pls. 1:14–18; 6:26; see Fig. 31C–H) described *Aptychus rugosus* from Köpinge, his specimens encompassing both Sharpe's *rugosus* (Pl. 1:14–18; Fig. 31C–E, G–H herein) and Hébert's *insignis* (Pl. 6:26; Fig. 31F herein). Moberg

also introduced a new taxon, *Aptychus flexus*, for material from Köpinge (Pl. 1:19; Fig. 31B herein), also referring to his new species the original of Schlüter (1876, Pl. 40:8), a further specimen from Köpinge. Moberg (1885, p. 41) also regarded *Aptychus stobaei* of Lundgren (1874, p. 73, Pl. 3:14–16) as a synonym of *rugosus* of Sharpe, and *leptophyllus* of Sharpe (1857, p. 57, Pl. 24:8–9) as a possible synonym. Trauth (1927) renamed the original of Schlüter (1876, Pl. 39:16) *Rugaptychus knorrianus*.

The only assemblage of *Rugaptychus* illustrated to date is that from the Campanian of Folx – les Caves, Brabant, Belgium (de Grossouvre 1908, Pl. 10:7–13; Kennedy 1987, p. 192, Pl. 16:1–22), which includes fragments that match in morphology *rugosus*, *insignis*, and *flexus* of previous authors, suggesting that these may be jaws of but a single *Baculites* species.

Superfamily Scaphitaceae Gill, 1871
Family Scaphitidae Gill, 1871
Subfamily Scaphitinae Gill, 1871

Genus and subgenus *Scaphites* Parkinson, 1811

Type species. – *Scaphites equalis* J. Sowerby, 1813, p. 53, Pl. 18:1–3, by the subsequent designation of Meek (1876, p. 413).

Scaphites (*Scaphites*) *hippocrepis* (DeKay, 1828) form III of Cobban, 1969

Fig. 32

Synonymy. – □1828 *Ammonites hippocrepis* – DeKay, p. 273, Pl. 5:5. □1885 *Scaphites* cf. *aquisgranensis* Schlüter – Moberg, p. 26, Pl. 3:2. □1969 *Scaphites hippocrepis* (DeKay) III – Cobban, p. 21, Pls. 3:1–25, 4:35–49, 5:36–40; Figs. 2, 4, 10–11 (with full synonymy). □1975 *Scaphites hippocrepis* (DeKay) – Schmid & Ernst, p. 332, Pl. 1:1–2; Fig. 1. □1993b *Scaphites* (*Scaphites*) *hippocrepis* (DeKay) – Kennedy & Cobban, p. 845, Figs. 15.1, 17.1–32.

Type. – Neotype, designated by Kennedy (1986a, p. 118) is no. 19483 in the collections of the Academy of Natural Sciences of Philadelphia, the holotype of *Scaphites cuvieri* Morton (1829, p. 109, Pl. 71:1) from the deep cut of the Chesapeake and Delaware Canal, USA.

Discussion. – Moberg (1885, p. 26, Pl. 3:2) described and illustrated as *Scaphites* cf. *aquisgranensis* Schlüter, a microconch from the Kåseberga. This specimen is RM 6327a (Fig. 32). It is badly crushed and distorted, and only 22 mm long. The spire is 16 mm in diameter with coarse ribs, the body chamber has flattened, subparallel flanks, weak umbilical and six strong ventrolateral clavi, with well-developed ventral ribs, all of which characters suggest form III of *hippocrepis* as recognized by Cobban (1969), or possibly a transition to his form II.

Occurrence. – *Scaphites* (*S.*) *hippocrepis* III is widespread in the Lower Campanian; in the United States it is known from Montana, Wyoming, South Dakota, Utah, Colorado and New Mexico, Travis and Lamar Counties in Texas, perhaps Russel County in Alabama, and is common in New Jersey and Delaware. In Europe, *S. hippocrepis* III or passage forms to *S. hippocrepis* II are known from Aquitaine and Alpes-Maritimes in France, southern England, Germany, Belgium, The Netherlands, plus the present record from southern Sweden. The successive subspecies mark a series of Lower Campanian Zones in the United States; in Europe, Schmid and Ernst (1973) recorded it

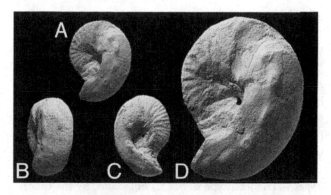

Fig. 32. □A–D. *Scaphites* (*Scaphites*) *hippocrepis* (DeKay, 1828) form III of Cobban (1969). RM Mo 6327a, the original of Moberg (1885, Pl. 3:2), from the lower Lower Campanian of Kåseberga. A–C ×1; D ×2.

from the upper Lower Campanian *conica/gracilis* Zone (=*conica/papillosa* Zone) and *gracilis/mucronata* Zone, and noted records down to *lingua/quadrata* Zone of the lower Lower Campanian (Fig. 2).

Scaphites (*Scaphites*) *binodosus* Roemer, 1841

Synonymy. – □1841 *Scaphites binodosus* – Roemer, p. 90, Pl. 13:6. □1841 *Scaphites inflatus* – Roemer, p. 90, Pl. 14:3. □1865 *Scaphites binodosus* Roemer – Roemer, p. 197, Pl. 32:6. □1872 *Scaphites binodosus* A. Röm. – Schlüter, p. 79, Pl. 24:4–6. □1872 *Scaphites inflatus* A. Röm. – Schlüter, p. 78, Pls. 24:1–3; 27:1. □ *non* 1897 *Scaphites binodosus* Röm. – Fritsch, p. 36, Fig. 20 (=*S. kieslingswaldensis* Langenhan & Grundey). □?1899 *Scaphites binodosus* Ad. Röm. – Schlüter, p. 414. □1902 *Scaphites inflatus* Roemer – Ravn, p. 252. □1902 *Scaphites binodosus* Roemer – Ravn, p. 253. □1905 *Scaphites inflatus* Roemer – Wegner, p. 211. □1905 *Scaphites binodosus* Roemer – Wegner, p. 211. □ *non* 1906 *Scaphites binodosus* A. Roemer – Müller & Wollemann, p. 16, Pls. 9:4–6; 10:4 (=*S. fisheri* Riedel, 1931). □1911 *Hoploscaphites inflatus* Römer – Nowak, p. 565. □1915 *Scaphites binodosus* A. Roemer – Frech, p. 560, Fig. 6. □1915 *Scaphites inflatus* F.A. Roemer – Frech, p. 561, Fig. 8. □1920 *Scaphites binodosus* A. Roemer – Köplitz, p. 67, Fig. 21. □1920 *Scaphites inflatus* A. Roemer – Köplitz, p. 68, Figs. 22–23. □1925 *Scaphites inflatus* Roemer – Diener, p. 199. □1925 *Discoscaphites binodosus* Roemer – Diener, p. 209. □1930 Ammonite spec. 1 – Hägg, p. 60, Pl. 5:9. □1931 *Scaphites binodosus* A. Röm. – Riedel, p. 700 (*pars*). □1935 *Scaphites binodosus* A. Roemer – Hägg, p. 60. □1943 *Scaphites* (*Discoscaphites*) *binodosus* A. Roemer – Hägg, p. 78, figure on p. 79. □1947 *Scaphites* (*Discoscaphites*) *binodosus* A. Roemer – Hägg, p. 94. □?1951 *Discoscaphites* cf. *binodosus* (Roemer) – Mikhailov, p. 95, Pl.

9:45. □1986a *Scaphites binodosus* (Roemer) – Kennedy, p. 116, Fig. 39A–E. □1987 *Scaphites binodosus* Roemer – Wright & Kennedy *in* Smith, Pl. 38:7, 11.

Name of the species. – Kennedy (1986a, p. 116) stated that he believed *S.* (*S.*) *binodosus* Roemer, 1841 (p. 90, Pl. 13:6) to be the microconch, and *S.* (*S.*) *inflatus* Roemer, 1841 (p. 90, Pl. 14:3) the macroconch of a single species, for which he selected the name *binodosus*.

Types. – The lectotype of *S.* (*S.*) *binodosus*, by the subsequent designation of Kennedy (1986a, p. 116), is the original of Roemer (1841, Pl. 13:6), refigured by Frech (1915, Fig. 6), from Dülmen, Westphalia, in the collections of the Geological Museum, Wrocław, Poland. Frech (1915, Fig. 8) also refigured the original of *Scaphites inflatus* Roemer, 1841, Pl. 141:3, but this specimen, from Dülmen, is now lost.

Discussion. – Hägg (1943, figure on p. 79) illustrated the venter of a typical individual from Ivö Klack, which we have not traced.

Occurrence. – Lower Campanian, with the record from Ivö Klack confirmed here; Hägg (1935) notes other localities. The species is also known from the Münster Basin and Braunschweig, Germany, Yorkshire, England, and Ukraine.

Scaphites (*Scaphites*) spp.

Discussion. – The *Ammonites* spec. 1 of Hägg (1930, p. 60, Pl. 5:9) from Eriksdal is the flank of a large, specifically indeterminate *Scaphites* of the group of *S.* (*S.*) *kieslingswaldensis fisheri* Riedel, 1931 – *S.* (*S.*) *binodosus* Roemer, 1841, but is too fragmentary for precise determination. The *Ammonites* spec. 3 of Hägg (1930, p. 61, Pl. 5:11), also from Eriksdal (LO 3156) is part of the body chamber and final hook of a *Scaphites* (*Scaphites*) with a maximum preserved whorl height of only 6 mm.

Genus *Trachyscaphites* Cobban & Scott, 1964

Type species. – *Trachyscaphites redbirdensis* Cobban & Scott, 1964, p. E7, Pl. 1:1–7; Fig. 3, by original designation.

Trachyscaphites spiniger spiniger (Schlüter, 1872)

Figs. 33–37

Synonymy. – □1841 *Scaphites pulcherrimus* – Roemer, p. 91 (*pars*), Pl. 14:4. □1872 *Scaphites spiniger* – Schlüter, p. 82, Pl. 25:1–7. □1885 *Scaphites spiniger* Schlüter – Moberg, p. 28, Pl. 3:6–8. □1889 *Scaphites spiniger* Schlüter – Griepenkerl, p. 405. □1894 *Scaphites spiniger* Schlüter – de Grossouvre, p. 252. □ *non* 1908 *Scaphites* cf. *spiniger* Schlüter – de Grossouvre, p. 38, Pl. 10:6; Fig. 13 [=*Hoploscaphites constrictus* (J. Sowerby)]. □1915 *Scaphites spiniger* Schlüter – Frech, p. 564, Fig. 13. □1916 *Acanthoscaphites spiniger* Schlüt. – Nowak, p. 67, figure opposite p. 66. □1925 *Scaphites* (*Acanthoscaphites*) *spiniger* Schlüter – Diener, p. 207. □1927 *Acanthoscaphites spiniger* (Schlüter) – Reeside, p. 34. □1951 *Acanthoscaphites spiniger* (Schlüter) – Mikhailov, p. 100, Pl. 19:92. □1952 *Scaphites spiniger* Schlüter – Basse, p. 612. □1954 *Scaphites* (*Acanthoscaphites*) *spiniger* Schlüter – Hägg, p. 58. □1954 *Scaphites* sp. 2 – Hägg, p. 58, Pl. 9:97. □1964 *Trachyscaphites spiniger* (Schlüter) – Cobban & Scott, p. E8. □1974 *Trachyscaphites spiniger* (Schlüter) – Naidin, p. 171, Pl. 59:2. □1975 *Scaphites spiniger* Schlüter – Schmid & Ernst, p. 330, Pls. 2:1–4; 3:3. □1976 *Trachyscaphites spiniger spiniger* (Schlüter) – Atabekian & Khakhimov, p. 66, Pls. 9:1–4; 12:2, 4. □1980 *Trachyscaphites spiniger* (Schlüter) – Błaskiewicz, p. 30, Pl. 13:1–3, 5–7. □1980 *Trachyscaphites spiniger posterior* Błaskiewicz, p. 31, Pls. 13:4, 14:1–7; 15:2–3; 30:2. □1986a *Trachyscaphites spiniger* (Schlüter) – Kennedy, p. 130, Pl. 22:4; Fig. 42. □1992 *Trachyscaphites spiniger spiniger* (Schlüter) – Cobban & Kennedy, p. 86, Pls. 1:2–3; 7:1–2, 5, 9; 8:1–9; Fig. 4A. □1993 *Trachyscaphites* cf. *spiniger* (Schlüter) – Kennedy, p. 113, Pl. 7:13.

Types. – Lectotype, by subsequent designation of Błaskiewicz (1980, p. 31), is the original of Schlüter (1872, Pl. 25:1–3), a macroconch, from the Upper Campanian of Darup, Westphalia, an unregistered specimen in the Institut für Paläontologie der Universität, Bonn (Fig. 35A–D). Paralectotypes are IPB 61a, the original of Schlüter (1872, Pl. 25:4), from the Hügelgruppe of Haldem, an adult microconch (Kennedy 1986a, Fig. 42A–B), and IPB 61b, from the same horizon and locality, the original of Schlüter (1872, Pl. 25:6) (Kennedy 1986a, Fig. 42F).

Fig. 33. □A–K. *Trachyscaphites spiniger spiniger* (Schlüter, 1872), probably middle Upper Campanian, Köpinge. □A. LO 7137t, microconch. □B–C. LO 7138t, macroconch. □D–E. LO 7139t, macroconch. □F–G. LO 727t, the original of Moberg (1885, Pl. 3:7). □H–K. LO 7140t, small macroconch. All ×1.

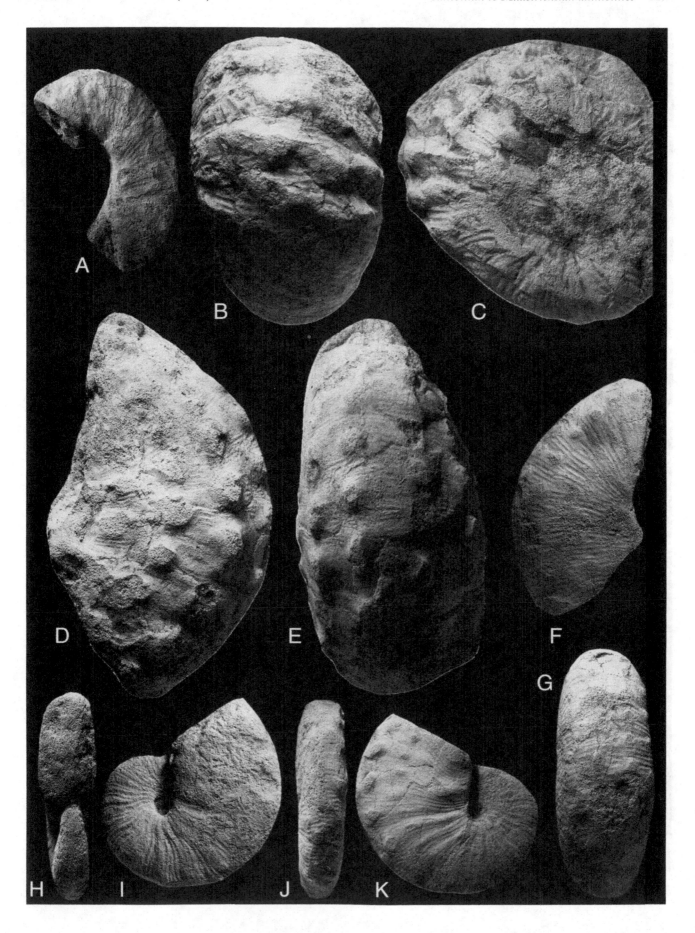

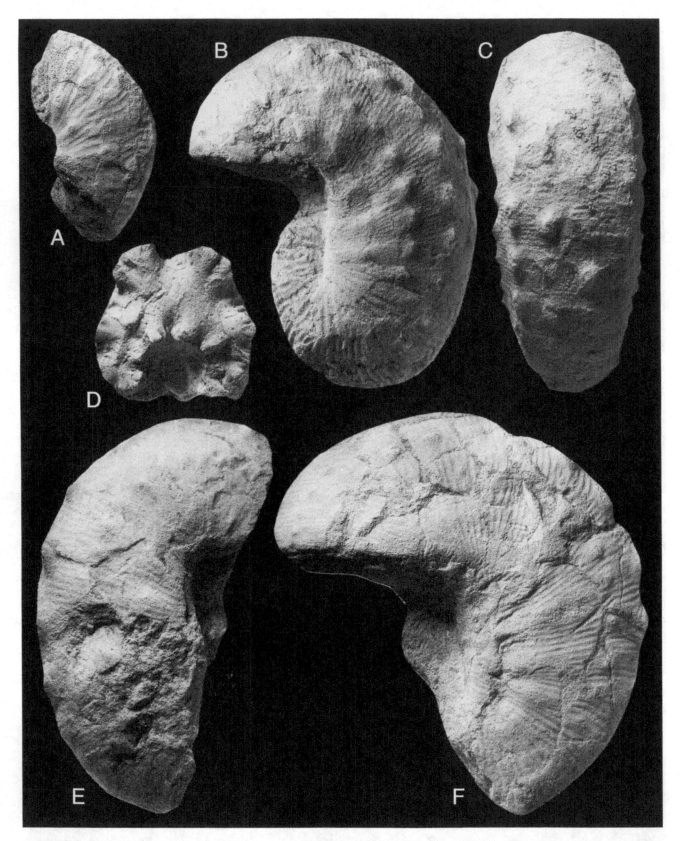

Fig. 34. □A–F. *Trachyscaphites spiniger spiniger* (Schlüter, 1872), probably middle Upper Campanian, Köpinge. □A. LO 7141t. □B–C. LO726t, the original of Moberg (1885, Pl. 3:6), macroconch. □D. LO 728t, the original of Moberg (1885, Pl. 3:8), a macroconch fragment. □E–F. LO 7142t. All ×1.

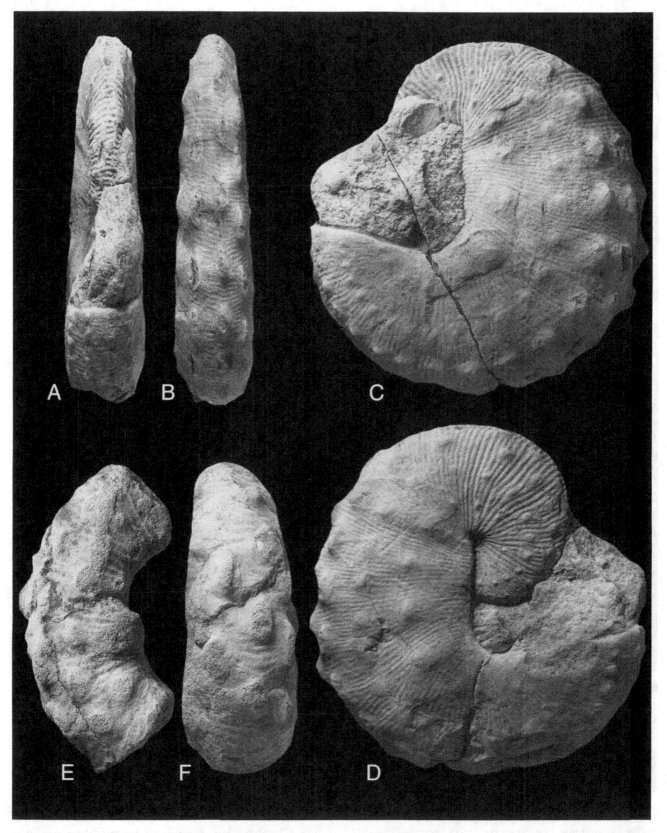

Fig. 35. □A–F. *Trachyscaphites spiniger spiniger* (Schlüter, 1872). □A–D. Lectotype, IPB Collections, the original of Schlüter (1872, Pl. 25:1–3), from the Upper Campanian of Darup, Westphalia, Germany. □E–F. LO 7143t, Köpinge, probably middle Upper Campanian. All ×1.

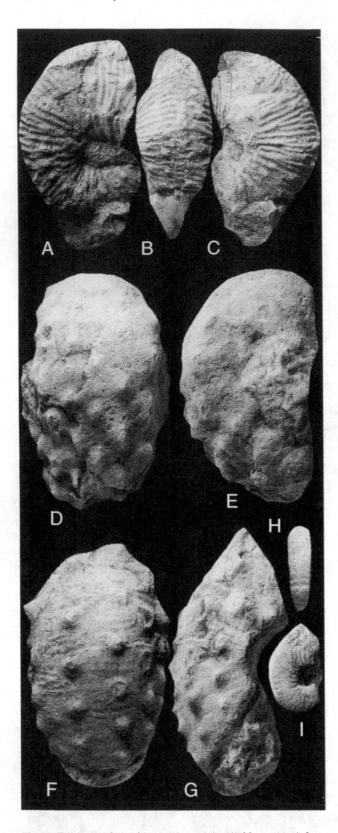

Fig. 36. □A–G. *Trachyscaphites spiniger spiniger* (Schlüter, 1872) from the Köpinge sandstone, probably middle Upper Campanian. □A–C. RM unregistered, the original of *Scaphites* sp. 2 of Hägg (1954, Pl. 9:97), Fredriksberg. □D–E. LO 7144t, Köpinge. □F–G. LO 7145t, Köpinge. □H–I. LO 7146t, juvenile scaphitid, cf. *Trachyscaphites spiniger spiniger*, Köpinge. All ×1.

Description. – We studied 43 fragmentary specimens, mostly from Köpinge. The species is strongly dimorphic. Phragmocones are very involute, with a tiny umbilicus. The whorl section varies from compressed (Fig. 33H–K) to depressed (Fig. 34D). Numerous delicate, straight, prorsiradiate primary ribs arise singly or in pairs at the umbilical shoulder. They are generally straight and prorsiradiate, and may increase by branching and intercalation. They link to a small outer lateral tubercle either singly or in pairs, or intercalate. The outer lateral tubercles give rise to 1–3 ribs, and these and the intercalated ribs loop, zigzag to, or intercalate between small conical inner ventrolateral tubercles, and connect in pairs or intercalate between small conical outer ventrolateral tubercles which alternate on either side of the venter; ribs loop, alternate or zig-zag between the outer ventrolateral tubercles (Figs. 33H–K, 36A–C). Some phragmocones develop small umbilicolateral/inner lateral tubercles at the adapertural end (Fig. 36A–C).

Microconch body chambers (Fig. 33A) are slender, with a concave umbilical wall and concave profile to the umbilical seam, not occluding the umbilicus of the phragmocone. Whorl sections vary from compressed to depressed. Delicate ribs arise at the umbilical seam and are concave across the umbilical wall. They link in groups at, or intercalate between seven coarse umbilical bullae, connected by delicate looped ribs to up to nine outer lateral tubercles, with other delicate ribs intercalated between. There are 9–11 inner and outer ventrolateral clavi that alternate and are linked by looped and zig-zag ribs, with delicate ribs intercalating. There is some variation in the relative strength of tubercle rows, and tuberculation may be strong or weak and may decline towards the adult aperture. The largest microconch in the collection is an estimated 60 mm long.

Macroconchs (Figs. 33B–K, 34A–C, E–F) have phragmocones from 35 mm (Fig. 33H–K) to 50 mm diameter (Figs. 33B–C, 34B–C), and are both compressed and slightly depressed, as in the only undeformed fragment (Fig. 34D). The largest complete macroconch is 115 mm long. The body chamber has high whorls compared with that of the microconch, with a convex umbilical wall and a straight profile to the umbilical seam that occludes and conceals part of the umbilicus of the phragmocone (Figs. 33I–K, 34B). The innermost row of tubercles is displaced out to an umbilicolateral position rather than being perched on the umbilical shoulder. Outer lateral, inner and outer ventrolateral tubercles are coarse, and linked by and separated by delicate ribs. There are both coarsely (Figs. 33B–C, D–E, 35E–F) and finely (Figs. 33H–K, 34B–C) tuberculate specimens; tuberculation weakens markedly towards the adult aperture (Figs. 34B–C, E–F), especially the umbilicolateral and outer lateral rows (Fig. 33F–G).

Fig. 37. Trachyscaphites spiniger spiniger (Schlüter, 1872). The holotype of *Trachyscaphites spiniger posterior* Błaskiewicz, 1980, the original of Błaskiewicz (1980, Pl. 14:5–7), no. 1,310.11.10 in the Institute of Geology Collections, Warsaw, from the Upper Campanian of Sulejów, Poland. All ×1.

Discussion. – The type material of *Trachyscaphites spiniger* comes from localities in the Münster Basin and from Haldem to the north. They have undergone substantial *post-mortem* crushing and their very compressed whorl section is in part illusory; the present material and that from sideritic concretions described by Cobban & Kennedy (1992) shows most specimens to have slightly depressed whorls, although compressed individuals also occur.

Trachyscaphites spiniger posterior Błaskiewicz, 1980 (p. 31, Pls. 13:4; 14:1–7; 15:2–3; 30:2), from the Upper Campanian of the Vistula Valley, Poland, was differentiated from the nominotypical subspecies because of the 'smaller number of ribs running between the tubercles of the same row on the exposed part of normal spiral and the presence of lateroumbilical tuberculation on earlier sectors of the exposed, normal spiral. It also differs on the whole in a smaller degree of freeing the shaft from phragmocone and in a frequent lack of ribs between the tubercles on the same row or shaft.' The holotype is shown in Fig. 37; we regard the name as a synonym of *spiniger sensu stricto*. In contrast, *Trachyscaphites spiniger porchi* (Adkins, 1929) (p. 205, Pl. 5:1–3), of which *Scaphites aricki* (Adkins, 1929) (p. 206, Pl. 5:7–8) is a synonym (see Cobban & Scott 1964, p. E10, Pls. 2:1–23; 3:1–11; Fig. 4), differs from the nominotypical subspecies in having fewer tubercles in all rows on the body chamber and generally lacking the dense ribbing so well-displayed by the present specimens. *Trachyscaphites spiniger levantinensis* Lewy,

1969 (p. 132, Pl. 4:1), from the Upper Campanian of Israel, is based on a microconch and is a synonym of *porchi*.

Trachyscaphites pulcherrimus (Roemer, 1841) (see revision in Kennedy & Summesberger 1984, p. 171, Pls. 11:1–2, 10–22, 13:2–6) is easily distinguished from *T. spiniger* by the presence of five rows of flank tubercles as well as a siphonal row in some individuals.

Occurrence. – Upper Campanian of Germany, The Netherlands, Belgium(?), Köpinge, Tosterup and Fredriksberg, Sweden; Ukraine, Armenia, Turkmenia, Poland; the species is restricted to the lower Upper Campanian where precisely dated. In the United States it is best known from the Ozan Formation in Fannin County, Texas.

Genus *Hoploscaphites* Nowak, 1911

Type species. – *Ammonites constrictus* – J. Sowerby, 1817, p. 189, Pl. A1, by original designation.

Hoploscaphites ikorfatensis Birkelund, 1965

Fig. 38

Synonymy. – □1872 *Scaphites Römeri* d'Orbigny – Schlüter, p. 89, (*pars*), Pl. 27:1–3, *non* 4 (=*Hoploscaphites*

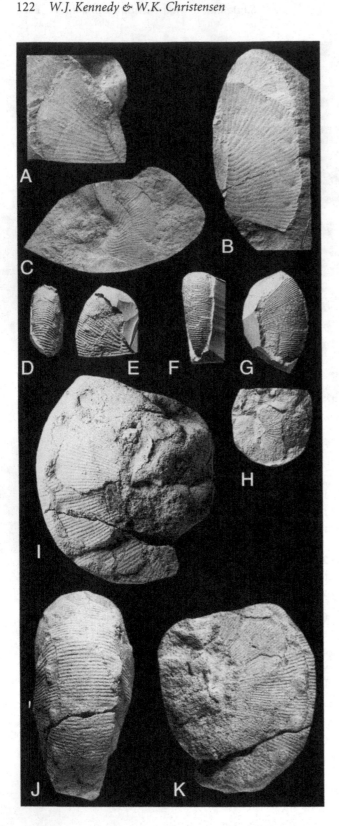

greenlandicus Donovan). □1885 *Scaphites Roemeri* d'Orbigny – Moberg, p. 29, Pl. 3:9. □1911 *Acanthoscaphites Römeri* d'Orbigny – Nowak, p. 585. □1916 *Acanthoscaphites Römeri* d'Orbigny – Nowak, p. 62. □1953 *Scaphites (Acanthoscaphites) roemeri* d'Orbigny – Ødum, p. 25, Pl. 2:2. □1954 *Scaphites (Acanthoscaphites) roemeri* d'Orbigny – Hägg, p. 57. □1954 *Scaphites greenlandicus* Donovan – Donovan, p. 8 (*pars*), Pl. 2:4; *non* Pl. 2:5, Fig. 1 (=*H. greenlandicus*). □1965 *Scaphites (Hoploscaphites) ikorfatensis* – Birkelund, p. 102, Pls. 24:1–4; 25:1–2; 26:1; Figs. 59, 93, 121 (3). □1965 *Scaphites (Hoploscaphites) ravni* – Birkelund, p. 106 (*pars*), Pls. 26:2–4; 27:1–4; ? *non* Pl. 28:1; *non* Pl. 29:1 (=*H. greenlandicus*); Figs. 95–96, ? 97, 100 (*pars*), 121 (4), ? *non* 121 (5).

Name of the species. – We regard *ikorfatensis* of Birkelund (1965) as the macroconch, of which co-occurring specimens of *ravni* of Birkelund (1965) (including the holotype) is the microconch. As first revising authors we select the name *ikorfatensis* for the species for those who accept this view.

Type. – Holotype is MGUH 9815, the original of Birkelund (1965, Pl. 26:1) from the Upper Campanian of Brudkløft at Ikorfat, West Greenland, 550 m above sea level. The holotype of *H. ravni* is MGUH 1819, the original of Birkelund (1965, Pl. 27:1, Fig. 95), from the same horizon and locality.

Description. – A phragmocone from Köpinge is 17 mm in diameter with the adapical part of the body chamber preserved (Fig. 38H). It is ornamented by dense prorsiradiate branching and intercalated ribs. A fragment from this locality with a whorl height of 17 mm (Fig. 38D–E) has delicate concave ribs at the umbilical shoulder that sweep forwards across to the mid-flank, where they increase by branching and intercalation and pass across the venter in a very shallow convexity. There are three delicate ventral tubercles on the fragment, which link up to three ribs, with four or more ribs intercalated between. A further fragment may be from close to the adult aperture, with an estimated whorl height of 18 mm (Fig. 38F–G). It has delicate ribs and lacks tubercles. Much larger is Moberg's specimen from Tosterup (1885, Pl. 3:9), LO 729t (Fig. 38I–K herein). This is a fragment of a body chamber with a maximum preserved whorl height of 31 mm, lacking the umbilical region. The whorl section is compressed, with flattened, convergent flanks, narrowly rounded ventrolateral shoulders and a feebly convex venter. Ornament is of

Fig. 38. □A–K. *Hoploscaphites ikorfatensis* Birkelund, 1965. □A–B. SGU unregistered, the original of Ødum (1954, Pl. 1:7), silicone squeeze, Upper Campanian of Höllviken-1 borehole, depth 643.4 m. □C. SGU unregistered, the original of Ødum (1954, Pl. 2:2), silicone squeeze, Upper Campanian, Höllviken-1 borehole, depth 639.25 m. □D–E. MGUH 24159, silicone mould, Köpinge, probably middle Upper Campanian. □F–G. MGUH 24160, silicone mould, Köpinge, probably middle Upper Campanian. □H. MGUH 24161, silicone mould, Köpinge, probably middle Upper Campanian. □I–K. LO 729t, the original of Moberg (1885, Pl. 3:9), probably lower Upper Campanian, Tosterup. All ×1.

delicate, prorsiradiate, feebly flexuous ribs/lirae which appear to increase by branching low on the flank, coarsen across the flanks and are near-transverse on the venter. Up to seven ribs link at coarse ventral clavi, with a similar number of ribs intercalated between.

The crushed specimens from the Höllviken-1 borehole are shown as Fig. 38A–C, and differ in no significant respects from the better-preserved Köpinge material.

Discussion. – Birkelund (1965) studied several hundred well-preserved and precisely localized Upper Campanian *Hoploscaphites* from West Greenland and clarified the relationships of the northwest European *Scaphites roemeri* group of authors. She recognized three species. *H. ikorfatensis* was based on macroconchs, with a single row of three to ten ventrolateral tubercles on the body chamber, or rarely nodeless. *H. greenlandicus* comes from a higher horizon and is also based on macroconchs and generally lacks tubercles, or may have only a few ventrolateral tubercles on the body chamber. *Hoploscaphites ravni* is based on microconchs; the holotype is from the same level as *H. ikorfatensis*, but Birkelund also referred to it specimens that co-occur with *H. greenlandicus*. It was diagnosed as having a single row of 0–10 ventrolateral tubercles on the body chamber, the tubercles rarely extending onto the phragmocone. The co-occurring macro-and microconch assemblages represent a dimorphic pair in our view, with *ravni* of Birkelund encompassing microconchs of both *H. ikorfatensis* and *H. greenlandicus*.

The *Scaphites compressus* of Roemer, 1841, was regarded as a homonym of *Scaphites compressus* of d'Orbigny by Giebel (1849), who renamed it *Scaphites tuberculatus*, and also by d'Orbigny (1850) who, unaware of Giebel's action, renamed it *Scaphites roemeri*, to which species the present material was formerly referred. The date of *compressus* of Roemer is 1841, but although d'Orbigny (1850, p. 214) gives the date of his *compressus* as 1841 and accords it priority over *compressus* of Roemer, the careful analysis of Sherborn (1889) shows that livraisons 33–42 (pp. 431–662) of d'Orbigny (1840–1842) date from 1842, so that *compressus* of Roemer has priority. Kennedy (1984, p. 148) discusses the affinities of d'Orbigny's *compressus*, and Birkelund (1965) discusses those of *roemeri* as interpreted by Schlüter (1872). The dense ribbing that covers the adult phragmocone and body chamber of *Scaphites compressus* Roemer, 1841, is highly distinctive and, together with the stout whorl section and bituberculation, distinguish it from the present species and other Old-World scaphites. It is a *Jeletzkytes* and resembles *J. nodosus* (Owen, 1852) (p. 581, Pl. 8:4) in general arrangement of ribs and tubercles, but the ribbing is finer even than that of the holotype of Owen's species, and the tubercles are weaker. *Scaphites elegans* Tate, 1865 (p. 37, Pl. 3:3; see Riccardi, 1983, Pl. 10:22–23) is no more than an uncrushed example of *Jeletzkytes compressus*.

Moberg (1885) recorded the Maastrichtian *Scaphites constrictus* from Köpinge; these records may well be based on fragments, or microconchs, of the present species.

Occurrence. – Upper Campanian of West Greenland, Köpinge, Tosterup, and the Höllviken-1 borehole, Sweden; the Münster Basin and Haldem, Westphalia, Germany.

Hoploscaphites constrictus (J. Sowerby, 1817)

Fig. 39A–E

Synonymy. – ☐1817 *Ammonites constrictus* – J. Sowerby, p. 189, Pl. A1. ☐1986b *Hoploscaphites constrictus* (J. Sowerby) – Kennedy, p. 64, Pls. 13:1–13; 16–24; 14:1–38; 15:1–31; Figs. 9, 11A–B (with full synonymy). ☐1993 *Hoploscaphites constrictus* (J. Sowerby) – Birkelund, p. 57, Pls. 14:1–7, 12; 15:1–14; 16:6–16; 17:5–23 (with additional synonymy). ☐1993 *Hoploscaphites constrictus* (J. Sowerby) – Kennedy, p. 113, Pl. 7:1–11, 14–16. ☐1993 *Hoploscaphites constrictus* (J. Sowerby) – Hancock & Kennedy, p. 166, Pl. 20:1–4 (with additional synonymy).

Types. – Lectotype, by the subsequent designation of Kennedy (1986b, p. 68), is BMNH C36733, the original of J. Sowerby (1817, Pl. A1); paralectotypes are BMNH C70645–70647, all from the Upper Maastrichtian Calcaire à *Baculites* of the Cotentin Peninsula, Manche, France.

Discussion. – Birkelund (1993) fully documented Swedish occurrences of this species and the closely related *H. tenuistriatus* (Kner, 1848). We illustrate here Moberg's figured specimens.

Occurrence. – *Hoploscaphites constrictus* ranges throughout almost all the Maastrichtian. At Kronsmoor in northern Germany, the first specimen appears 3.5–5.0 m above the base of the *Belemnella lanceolata* Zone, while at Stevns Klint in Denmark it is common in the topmost hardground of the Maastrichtian White Chalk, which is immediately overlain by the Palaeocene Fish Clay. It is known from the Biscay region of France and Spain, Petites Pyrénées (Haute Garonne), Tercis (Landes) and the Calcaire à *Baculites* of the Cotentin Peninsula (Manche) (all in France), the Nekum and Meerssen Chalks of the Maastricht region in Belgium and The Netherlands, Germany, Denmark, southern Sweden, Poland, Austria (Styria), the Czech Republic, Bulgaria, Ukraine, Donbass, Carpathians, Transcaspia, Kazakhstan, and Kopet Dag (Turkmenia).

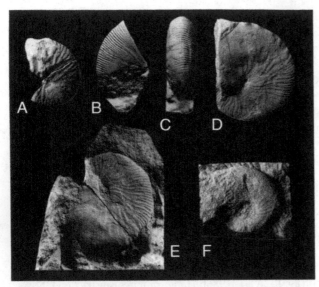

Fig. 39. □A–E. *Hoploscaphites constrictus* (J. Sowerby, 1817), upper Lower Maastrichtian, Ulricelund near Näsbyholm. □A. LO 725t, the original of Moberg (1885, Pl. 3:5). □B–C. LO 724t, the original of Moberg (1885, Pl. 3:4). □D–E. LO 723t, the original of Moberg (1885, Pl. 3:3). □F. *Hoploscaphites tenuistriatus* (Kner, 1848). SGU Collections, the original of Ødum (1954, Pl. 1:6), Upper Maastrichtian, Höllviken-1 borehole, depth 375.57 m. All ×1.

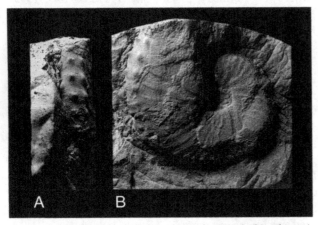

Fig. 40. *Acanthoscaphites* sp. SGU unregistered, original of *Scaphites tridens* Kner of Ødum (1954, Pl. 1:4), from the Maastrichtian of the Höllviken-1 borehole, depth 314.5 m. All ×1.

Hoploscaphites tenuistriatus (Kner, 1848)

Fig. 39F

Synonymy. – □1848 *Scaphites tenuistriatus* – Kner, p. 10, Pl. 1:5. □1987 *Hoploscaphites tenuistriatus* (Kner) – Kennedy, p. 210, Pl. 31:2–7 (with synonymy). □1993 *Hoploscaphites tenuistriatus* (Kner) – Birkelund, p. 59, Pl. 14:8–11, 13–16.

Type. – Kner's specimen from Lemberg (Lvov) (1848, Pl. 1:5) should be designated lectotype if still in existence.

Discussion. – Birkelund (1993) revised Danish occurrences. A juvenile from the Höllviken-1 borehole at a depth of 375.75 m is shown as Fig. 39F.

Occurrence. – High Lower to low Upper Maastrichtian. There are records from Denmark, Sweden, Germany, The Netherlands, Poland, the Czech Republic, Ukraine and southern Russia.

Genus *Acanthoscaphites* Nowak, 1911

Type species. – *Scaphites tridens* Kner, 1848, p. 10, Pl. 2:1, by subsequent designation by Diener (1925, p. 205).

Acanthoscaphites sp.

Fig. 40

Synonymy. – □1953 *Scaphites* (*Acanthoscaphites*) *tridens* Kner – Ødum, p. 24, Pl. 1:4.

Description. – The specimen is a composite mould, 37 mm in diameter, and may include part of the body chamber. Coiling is involute, the umbilicus small, shallow, with a flattened umbilical wall and narrowly rounded umbilical shoulder. Widely-separated narrow ribs are concave on the innermost flank but become straight and prorsiradiate and split into pairs of delicate riblets or lirae on the inner flank with numerous delicate growth lines between. Ribs are feebly concave on the outer flank and terminate in delicate conical ventral tubercles. The ventral view is partially concealed (Fig. 40A), but delicate growth lines link to a second row of tubercles, while traces of a third row, alternating in position with the first row are also visible. These tubercle rows may be inner and outer ventrolateral or ventral, siphonal and ventral; it cannot be determined if there were three or five rows of tubercles across the venter.

Discussion. – This specimen may be a microconch of *Acanthoscaphites* close to *A. varians* (Łopuski, 1911) (pp. 120, 137, Pl. 4:1–3; Fig. 1; see revision in Birkelund 1993, p. 56; Pls. 9:3–7; 10:2–3), but uncertainty as to ventral tuberculation precludes close comparisons.

Occurrence. – Upper Maastrichtian, Höllviken-1 borehole, 314.5 m.

Acknowledgements. – We thank G. Kjellström, Geological Survey of Sweden, Uppsala, S. Laufeld, formerly at the Geological Survey of Sweden, Uppsala, J. Bergström, Swedish Museum of Natural History, Section of Palaeozoology, Stockholm, and K. Larsson, Geological Institute, University of Lund, for placing ammonites at our disposal. We also thank H.C. Klinger (Cape Town), A.V. Dhondt (Brussels), M. Sander (Bonn), A. Błaskiewicz (Warsaw), W.A. Cobban (Denver), U. Kaplan (Gütersloh), U. Scheer (Essen), the late H. Jaeger (Berlin), T. Stuart (Norwich), H. Gauthier (Paris), and M. Bilotte (Toulouse) for advice, discussion, and access to collections in their care. P. Bengtson (Heidelberg) reviewed the manuscript. Kennedy acknowledges the financial support of the Natural Environment Research Council (U.K.), and the technical assistance of the staff of the Geological Collections, University Museum, Oxford and Department of Earth Sciences, Oxford. Thomas Bredsdorff, Geological Museum, Copenhagen prepared the photographic illustrations.

References

Adkins, W.S. 1929: Some Upper Cretaceous Taylor ammonites from Texas. *University of Texas Bulletin 2901*, 203–211, 220–223.

Anderson, F.M. 1958: Upper Cretaceous of the Pacific Coast. *Geological Society of America Memoir 71*, 378 pp.

Arnold, H. 1964: Die Erforschung der westfälischen Kreide und zur Definition der Oberkreidestufen und -zonen. *Fortschritte in der Geologie von Rheinland und Westfalen 7*, 1–14.

Atabekian, A.A. & Khakhimov, F.Kh. 1976: *Kampanskie i maastrikhtskie ammonity srednej Azii.* [*Campanian and Maastrichtian ammonites from central Asia.*] 146 pp. Donish, Dushanbe. [In Russian.]

Baily, W.H. 1855: Descriptions of some Cretaceous fossils from South Africa. *Quarterly Journal of the Geological Society of London 11*, 454–465.

Basse, E. 1952: Ammonoïdes. *In* Piveteau, J. (ed.): *Traité de Paléontologie 2*, 522–555, 581–688. Masson, Paris.

Binkhorst, J.T. 1861: *Monographie des Gastropodes et des Céphalopodes de la Craie Supérieur du Limbourg.* 17+83 pp. (gastropods)+44 pp (cephalopods), Muquardt, Maastricht; Muller Frères, Brussels.

Birkelund, T. 1957: Upper Cretaceous belemnites from Denmark. *Biologiske Skrifter. Det Kongelige Danske Videnskabernes Selskab 9*. 69 pp.

Birkelund, T. 1965: Ammonites from the Upper Cretaceous of West Greenland. *Meddelelser om Grønland 179*. 192 pp.

Birkelund, T. 1973: A note on *Pseudopuzosia* sp. from Särdal. *In* Bergström, J., Christensen, W.K., Johansson, C. & Norling, E.: An extension of Upper Cretaceous rocks to the Swedish west coast at Särdal. *Bulletin of the Geological Society of Denmark 22*, 141–142.

Birkelund, T. 1993: Ammonites from the Maastrichtian White Chalk in Denmark. *Bulletin of the Geological Society of Denmark 40*, 33–81.

Birkelund, T. & Bromley, R.G. 1979: *Hauericeras* cf. *pseudogardeni* in the Upper Cretaceous of Ignaberga, Sweden. *Geologiska Föreningens i Stockholm Förhandlingar 101*, 173–176.

Błaskiewicz, A. 1980: Campanian and Maastrichtian ammonites of the Middle Vistula River Valley, Poland: a stratigraphic–paleontologic study. *Prace Instytutu Geologicznego 92*. 63 pp.

Boule, M., Lemoine, P. & Thévenin, A. 1906–1907: Paléontologie de Madagascar III. Céphalopodes crétacés des environs de Diego–Suarez. *Annales de Paléontologie 1*, 173–192 [1–20]; *2*, 1–56 [21–76].

Brinkmann, R. 1935: Die Ammoniten der Gosau und des Flysch in den nördlichen Ostalpen. *Mitteilungen aus dem Geologischen Staatsinstitut im Hamburg 15*, 1–14.

Brotzen, F. 1945: De geologiska resultaten från borrningarne vid Höllviken. 1. Kritan. *Sveriges Geologiska Undersökning C465*. 64 pp.

Brotzen, F. 1958: [Cretaceous.] *In* Magnusson, N.H. (ed.): *Lexique Stratigraphique International. 1. Europe. Fasc. 2c: Sweden.* Paris.

Brotzen, F. 1960: The Mesozoic of Scania, Southern Sweden. International Geological Conress, XXI Session, Norden 1960. *Guide to Excursions A21 and C16.* 15 pp.

Christensen, W.K. 1970: *Actinocamax primus* Arkhangelsky from Scania, Sweden. *Stockholm Contributions in Geology 21*, 69–76.

Christensen, W.K. 1973: The belemnites and their stratigraphic significance. *In* Bergström, J., Christensen, W.K., Johansson, C. & Norling, E.: An extension of Upper Cretaceous rocks to the Swedish west coast at Särdal. *Bulletin of the Geological Society of Denmark 22*, 113–140.

Christensen, W.K. 1975: Upper Cretaceous belemnites from the Kristianstad area in Scania. *Fossils and Strata 7*. 69 pp.

Christensen, W.K. 1985 [date of imprint 1984]: The Albian to Maastrichtian of southern Sweden and Bornholm, Denmark: a review. *Cretaceous Research 5*, 313–327.

Christensen, W.K. 1986: Upper Cretaceous belemnites from the Vomb Trough in Scania. *Sveriges Geologiska Undersökning Ca57.* 57 pp.

Christensen, W.K. 1993: Upper Cretaceous belemnitellids from the Båstad Basin, southern Sweden. *Geologiska Föreningens i Stockholm Förhandlingar 115*, 39–57.

Christensen, W.K. 1995: *Belemnitella* from the Upper Campanian and Lower Maastrichtian Chalk of Norfolk, England. *Special Papers in Palaeontology 51.* 84 pp.

Christensen, W.K. & Schulz, M.-G. 1976: First record of *Belemnellocamax balsvikensis* (Brotzen, 1960) from NW Germany. *Neues Jahrbuch für Geologie und Paläontologie, Monatshefte 9*, 522–531.

Christensen, W.K. & Schulz, M.-G. 1997: Coniacian and Santonian belemnite faunas from Bornholm, Denmark. *Fossils and Strata 44*, 1–73.

Cobban, W.A. 1962: *Baculites* from the lower part of the Pierre Shale and equivalent rocks in the Western Interior. *Journal of Paleontology 36*, 704–718.

Cobban, W.A. 1969: The Late Cretaceous ammonites *Scaphites leei* Reeside and *Scaphites hippocrepis* (DeKay) in the Western Interior of the United States. *United States Geological Survey Professional Paper 619*, 1–27.

Cobban, W.A. & Kennedy, W.J. 1992: Campanian *Trachyscaphites spiniger* ammonite fauna in north-east Texas. *Palaeontology 35*, 63–93.

Cobban, W.A. & Scott, G.R. 1964: Multinodose scaphitid cephalopods from the lower part of the Pierre Shale and equivalent rocks in the conterminous United States. *United States Geological Survey Professional Paper 483-E*, E1–E13.

Collignon, M. 1952: Ammonites néocrétacées du Menabe (Madagascar). II – Les Pachydiscidae. *Travaux du Bureau Géologique du Haut Commissariat de Madagascar et Dépendances 41.* 114 pp.

Collignon, M. 1955: Ammonites néocrétacées du Menabe (Madagascar). II. Les Pachydiscidae. *Annales Géologiques du Service des Mines de Madagascar 21.* 98 pp.

Collignon, M. 1961: Ammonites néocrétacées du Menabe (Madagascar). VII. Les Desmoceratidae. *Annales Géologiques du Service des Mines de Madagascar 31.* 115 pp.

Collignon, M. 1971: Atlas des fossiles charactéristiques de Madagascar (Ammonites) XVII (Maestrichtien). *Service Géologique, Tananarive.* 44 pp.

Collignon, M. 1983: Les faunes d'ammonites du Santonien. *Documents des Laboratoires de Géologie de Lyon H.S. 6 (for 1981)*, 184–224.

DeKay, J.E. 1828: Report on several fossil multilocular shells from the state of Delaware: with observations on a second specimen of the new fossil genus *Eurypterus*. *Annals of the Lyceum of Natural History of New York 2*, 273–279.

Desmarest, A.G. 1817: Mémoire sur deux genres coquilles fossiles cloisonnées et à siphon. *Journal de Physique de Chimie, d'Histoire Naturelles et des Arts 85*, 42–51.

Diener, C. 1925: Ammonoidea neocretacea. *Fossilium Catalogus (1: Animalia) 29.* 244 pp.

Donovan, D.T. 1954: Upper Cretaceous fossils from Traill and Geographical Society Øer, East Greenland. *Meddelelser om Grønland 72.* 33 pp.

Douvillé, H. 1890: Sur la classification des Cératites de la Craie. *Bulletin de la Société Géologique de France (3) 18*, 275–292.

Dujardin, F. 1837: Mémoire sur les couches du sol en Touraine et déscription des coquilles et la craie et des faluns. *Mémoire de la Société Géologique de France 2*, 211–311.

Erlström, M. & Gabrielson, J. 1992: Petrology, fossil composition and depositional history of the Ignaberga limestone, Kristianstad Basin, Scania. *Sveriges Geologiska Undersökning Ca80.* 30 pp.

Ernst, G. 1964: Ontogenie, Phylogenie und Stratigraphie der Belemnitengattung *Gonioteuthis* Bayle aus dem nordwestdeutschen Santon/Campan. *Fortschritte in der Geologie von Rheinland und Westfalen 7*, 113–174.

Ernst, G., Schmid, F. & Klischies, G. 1979: Multistratigraphische Untersuchungen in der Oberkreide des Raumes Braunschweig–Hannover. *In* Wiedmann, J. (ed.): Aspekte der Kreide Europas. *International Union of Geological Sciences A6*, 11–46.

Forbes, E. 1846: Report on the fossil invertebrata from southern India, collected by Mr. Kaye and Mr. Cunliffe. *Transactions of the Geological Society of London (2) 7*, 97–174.

Frech, F. 1910: Geologische Beobachtungen im pontischen Gebirge. *Neues Jahrbuch für Mineralogie, Geologie und Paläontologie 1910 (1)*, 1–24.

Frech, F. 1915: Über *Scaphites*. 1. Die Bedeutung von *Scaphites* für die Gliederung der Oberkreide. *Centralblatt für Mineralogie, Geologie und Paläontologie 1915*, 553–568.

Fritsch, A. 1897: Studien im Gebiete der böhmischen Kreideformation. Palaeontologische Untersuchungen der einzelnen Schichten. VI. Die Chlomeker Schichten. *Archiv für die Naturwissenschaftliche Landesdurchforschung von Böhmen 10.* 84 pp.

Fritsch, A. & Kafka, J. 1887: *Die Crustaceen der böhmischen Kreideformation.* 53 pp. Selbstverlag, Prague.

Giebel, C.G. 1849: Sitzung am 14. November. *Jahresbericht des Naturwissenschaftlichen Vereins in Halle 1849,* 18–20.

Giers, R. 1964: Die Grossfauna der Mukronatenkreide (unteres Obercampan) im östlichen Münsterland. *Fortschritte in der Geologie von Rheinland und Westfalen 7,* 213–194.

Gill, T. 1871: Arrangement of the families of Mollusks. *Smithsonian Miscellaneous Collections 227.* 49 pp.

Griepenkerl, O. 1889: Die Versteinerungen der Senonen Kreide von Koenigslutter im Herzogthum Braunschweig. *Palaeontologische Abhandlungen 4,* 305–419 (3–116).

Grossouvre, A. de 1894 [date of imprint 1893]: Recherches sur la craie supérieur, 2, Paléontologie. Les ammonites de la craie supérieur. *Mémoires du Service de la Carte Géologique détaillée de la France.* 264 pp.

Grossouvre, A. de 1908: Description des ammonites du Crétacé supérieur du Limbourg belge et hollandais et du Hainault. *Mémoires du Musée Royal d'Histoire Naturelle de Belgique 4,* 1–39.

Hägg, R. 1930: Die Mollusken und Brachiopoden der schwedischen Kreide. 1. Eriksdal. *Sveriges Geologiska Undersökning C363.* 94 pp.

Hägg, R. 1935: Die Mollusken und Brachiopoden der schwedischen Kreide. 2. Kullemölla, Lyckås, Kåseberga und Gräsryd. *Sveriges Geologiska Undersökning C385.* 94 pp.

Hägg, R. 1943: *Scaphites (Discoscaphites) binodosus* A. Roemer från Blaksudden på Ivö. *Geologiska Föreningens i Stockholm Förhandlingar 65,* 78–79.

Hägg, R. 1947: Die Mollusken und Brachiopoden der schwedischen Kreide. 3. Das Kristianstadgebiet. *Sveriges Geologiska Undersökning C485.* 143 pp.

Hägg, R. 1954: Die Mollusken und Brachiopoden der schwedischen Kreide. 4. Die Mammillaten- und Mucronatenkreide des Ystadgebietes. *Sveriges Geologiska Undersökning C535.* 72 pp.

Hancock, J.M. & Kennedy, W.J. 1993: The high Cretaceous ammonite fauna from Tercis, Landes, France. *Bulletin de l'Institut Royal des Sciences Naturelles de Belgique, Sciences de la Terre 63,* 149–209.

Hauer, F. von 1858: Ueber die Cephalopoden der Gosauschichten. *Beiträge zur Paläontographie von Österreich 1,* 7–14.

Hauer, F. von 1866: Neue Cephalopoen aus dem Gosau-gebilden der Alpen. *Sitzungsberichte der Kaiserlichen Akademie der Wissenschaften in Wien, Mathematisch-naturwissenschaftliche Klasse 53,* 1–9.

Hébert, E. 1856: Tableau des fossiles de craie de Meudon et déscription de quelques espèces nouvelles. *Mémoire de la Société Géologique de France (2) 5 (2),* 345–374.

Henderson, R.A. & McNamara, K.A. 1985: Maastrichtian non-heteromorph ammonites from the Mira Formation, Western Australia. *Palaeontology 28,* 35–88.

Hyatt, A. 1889: Genesis of the Arietidae. *Smithsonian Contributions to Knowledge 673.* 239 pp.

Hyatt, A. 1894: Phylogeny of an acquired characteristic. *Proceedings of the American Philosophical Society 32,* 349–647.

Hyatt, A. 1900: Cephalopoda. *In* Zittel, K.A. von 1896–1900: *Textbook of Palaeontology,* 502–592. Macmillan, London. [English translation.]

Jagt, J.M.W. 1988: Some stratigraphical and faunal aspects of the Upper Cretaceous of southern Limburg (The Netherlands) and contiguous areas. *In* Streel, M. & Bless, M.J.M. (eds.): *The Chalk District of the Euregio Meuse-Rhine,* 25–39. Natuurhistorisch Museum, Maastricht, and Laboratoires de Paléontologie de l'Université d'Etat à Liège, Liège.

Jeletzky, J.A. 1948: Zur Kenntnis der Oberkreide der Dnepr-Donez-Senke und zum Vergleich der russischen borealen Oberkreide mit derjenigen Polens und Nordwesteuropas. *Geologiska Föreningens i Stockholm Förhandlingar 70,* 583–602.

Jeletzky, J.A. 1951: Die Stratigraphie und Belemnitenfauna des Obercampan und Maastricht Westfalens, Nordwestdeutschlands und Dänemarks, sowie einige allgemeine Gliederungs-Probleme der jüngeren borealen Oberkreide Eurasiens. *Beihefte Geologisches Jahrbuch 1.* 142 pp.

Jones, B. 1963: Upper Cretaceous (Campanian and Maastrichtian) ammonites from southern Alaska. *United States Geological Survey Professional Paper 432.* 55 pp.

Kaplan, U. & Kennedy, W.J. 1994: Ammoniten des Westfälischen Coniac. *Geologie und Paläontologie in Westfalen 31.* 155 pp.

Kayser, E. 1924: *Lehrbuch der Geologie 11, Geologischen Formationskunde.* 7 Ed. 657 pp. Ferdinand Enke, Stuttgart.

Kennedy, W.J. 1984: Systematic palaeontology and stratigraphical distribution of the ammonite faunas of the French Coniacian. *Special Papers in Palaeontology 31.* 160 pp.

Kennedy, W.J. 1986a: Campanian and Maastrichtian ammonites from northern Aquitaine, France. *Special Papers in Palaeontology 36.* 145 pp.

Kennedy, W.J. 1986b: The ammonite fauna of the Calcaire à *Baculites* (Upper Maastrichtian) of the Cotentin Peninsula (Manche, France). *Palaeontology 29,* 25–83.

Kennedy, W.J. 1987 [date of imprint 1986]: The ammonite faunas of the type Maastrichtian, with a review of *Ammonites colligatus* Binkhorst, 1861. *Bulletin de l'Institut Royal des Sciences Naturelles de Belgique, Sciences de la Terre 56,* 51–267.

Kennedy, W.J. 1993: Campanian and Maastrichtian ammonites from the Mons Basin (Belgium). *Bulletin de l'Institut Royal des Sciences Naturelles de Belgique, Sciences de la Terre 63,* 99–131.

Kennedy, W.J. & Christensen, W.K. 1991: Coniacian and Santonian ammonites from Bornholm, Denmark. *Bulletin of the Geological Society of Denmark 38,* 203–226.

Kennedy, W.J. & Christensen, W.K. 1993: Santonian ammonites from the Köpingsberg-1 borehole, Sweden. *Bulletin of the Geological Society of Denmark 40,* 149–156.

Kennedy, W.J. & Cobban, W.A. 1988: Mid-Turonian ammonite faunas from northern Mexico. *Geological Magazine 125,* 593–612.

Kennedy, W.J. & Cobban, W.A. 1993a: Ammonites from the Saratoga Chalk (Upper Cretaceous) of Arkansas, U.S.A. *Journal of Paleontology 67,* 404–434.

Kennedy, W.J. & Cobban, W.A. 1993b: Lower Campanian (Upper Cretaceous) ammonites from the Merchantville Formation of New Jersey, Maryland and Delaware. *Journal of Paleontology 67,* 828–849.

Kennedy, W.J. & Hancock, J.M. 1993: Upper Maastrichtian ammonites from the Marnes de Nay between Gan and Rébénacq (Pyrénées-Atlantique), France. *Géobios 26,* 575–594.

Kennedy, W.J., Hansotte, M., Bilotte, M. & Burnett, J. 1992: Ammonites and nannofossils from the Campanian of Nalzen (Ariège, France). *Géobios 25,* 263–278.

Kennedy, W.J. & Henderson, R.A. 1992: Heteromorph ammonites from the Upper Maastrichtian of Pondicherry, South India. *Palaeontology 35,* 693–731.

Kennedy, W.J. & Klinger, H.C. 1977: Cretaceous faunas from Zululand and Natal, South Africa. The ammonite family Tetragonitidae Hyatt, 1900. *Annals of the South African Museum 73:7,* 149–197.

Kennedy, W.J. & Summesberger, H. 1984: Upper Campanian ammonites from the Gschliefgraben (Ultrahelvetic, Upper Austria). *Beiträge zur Paläontologie von Österreich 11,* 149–206.

Kennedy, W.J. & Summesberger, H. 1987: Lower Maastrichtian ammonites from Nagoryany (Ukrainian SSR). *Beiträge zur Paläontologie von Österreich 13,* 25–78.

Keutgen, N. 1995: Late Campanian *Belemnitella* from Zeven Wegen Member (Gulpen Formation) at the CPL quarry (Haccourt, NE Belgium). *Second International Symposium on Cretaceous Stage Boundaries, Abstract Volume,* p. 176 only. Institut Royal des Sciences Naturelles de Belgique, Brussels.

Kilian, W. & Reboul, P. 1909: Les céphalopodes néocrétacés des îles Seymour et Snow Hill. *Wissenschaftliche Ergebnisse der schwedischen Südpolar-Expedition 3:6.* 75 pp.

Kner, R. 1848: Versteinerungen des Kreidemergels von Lemberg und seiner Umgebung. *Haidingers Naturwissenschaftliche Abhandlungen 2*. 42 pp.

[Köplitz, W. 1920: Über die Fauna des oberen Untersenon im Seppenrade – Dülmener Höhenzuge. Dissertation, Westfalia Wilhelms-Universität Münster. 78 pp.]

Kossmat, F. 1895–1898: Untersuchungen über die südindische Kreideformation. *Beiträge zur Paläontologie Österreich–Ungarns und des Orients 9*, 97–203 [1–107] (1895); *11*, 1–46 [108–153] (1897); *11*, 89–152 [154–217] (1898).

Lamarck, J.P.B.A. de M. de 1799: Prodome d'une nouvelle classification des coquilles. *Mémoire de la Société d'Histoire Naturelle de Paris (1799)*, 63–90.

Lamarck, J.P.B.A. de M. de 1801: *Systéme des Animaux sans vertèbres*. 432 pp. Déterville, Paris.

Lamarck, J.P.B.A. de M. de 1822: *Histoire naturelles des Animaux sans vertèbres 7*. 711 pp. Verdière, Paris.

Leonhard, R. 1897: Die Fauna der Kreideformation in Oberschlesien. *Palaeontographica 44*, 11–70.

Lewy, Z. 1969: Late Campanian heteromorph ammonites from southern Israel. *Israel Journal of Earth Sciences 18*, 109–135.

Łopuski, C. 1911: Przyczunki do znajmosci fauny kredowej gub, Lubelskiej. *Compte Rendu des Séances et la Société Scientifique de Varsovie 4*, 104–140.

Lundegren, A. 1932: Om förekomsten av Cenoman i Båstads-området och dess betydelse för dateringen av uppkomsten av Hallandsås. *Geologiska Föreningens i Stockholm Förhandlingar 54*, 500–504.

Lundegren, A. 1933: Köpingesandstenen i sydöstra Skåne. *Geologiska Föreningens i Stockholm Förhandlingar 55*, 163–165.

Lundegren, A. 1935: Die stratigraphischen Ergebnisse der Tiefbohrung bei Kullemölla im südöstlichen Schonen. *Sveriges Geologiska Undersökning C386*, 18 pp.

Lundgren, B. 1874: Om en *Comaster* och en Aptychus från Köpinge. *Öfversigt af Kongliga Vetenskaps-Akademiens Förhandlingar 3*, 61–73.

Lundgren, B. 1881: Om *Scaphites binodosus* Roem. från Kåseberga. *Öfversigt af Kongliga Vetenskaps-Akademiens Förhandlingar 10*, 23–28.

Mariani, E. 1898: Ammoniti del Senoniano Lombardo. *Memoire del Reale Instituto Lombardi di Scienze e Lettere, Classe di Scienze Matematiche e Naturali (3) 18*, 51–58 (1–18).

Marshall, P. 1926: The Upper Cretaceous ammonites of New Zealand. *Transactions of the New Zealand Institute 56*, 129–210.

Matsumoto, T. 1938: A biostratigraphic study on the Cretaceous deposits of the Naibuchi Valley, South Karahuto. *Proceedings of the Imperial Academy of Japan 14*, 190–194

Matsumoto, T. 1955: The bituberculate pachydiscids from Hokkaido and Saghalien. *Memoirs of the Faculty of Science, Kyushu University, Series D, Geology 5*, 153–184.

Matsumoto, T. 1959: Upper Cretaceous ammonites of California. Part 1. *Memoirs of the Faculty of Science, Kyushu University, Series D, Geology 8*, 91–171.

Matsumoto, T. 1979: Notes on *Lewesiceras* and *Nowakites* (pachydiscid ammonites) from the Cretaceous of Hokkaido. *Transactions and Proceedings of the Palaeontological Society of Japan, N.S. 113*, 30–44.

Matsumoto, T. & Miyauchi, T. 1984: Some Campanian ammonites from the Soya area. *Special Paper, Palaeontological Society of Japan 27*, 33–91.

Matsumoto, T. & Obata, I. 1955: Some Upper Cretaceous desmoceratids from Hokkaido and Saghalien. *Memoirs of the Faculty of Science, Kyushu University, Series D, Geology 5*, 119–151.

Matsumoto, T., Toshimitsu, S.A. & Kawashita, Y. 1990: On *Hauericeras* de Grossouvre, 1894, a Cretaceous ammonite. *Transactions and Proceedings of the Palaeontological Society of Japan, N.S. 158*, 439–458.

Meek, F.B. 1876: A report on the invertebrate Cretaceous and Tertiary fossils of the upper Missouri country. *In* Hayden, F.V.: *Report of the United States Geological Survey of the Territories 9*. 629 pp.

Mikhailov, N.P. 1951: Verkhnemelovye ammonity yuga evropejskoj chasti SSSR i ikh znachenie dlya zonal'noj stratigrafii. [Upper Creta-

ceous ammonites from the southern part of European Russia and their importance for zonal stratigraphy (Campanian, Maastrichtian).] *Trudy Instituta Geologicheskikh Nauk 129, Geology Series 50*, 143 pp. [In Russian.]

Moberg, J.C. 1885: Cephalopoderna i Sveriges kritsystem. II. Artsbeskrifning. *Sveriges Geologiska Undersökning C73*, 63 pp.

Moore, R.C. (ed.) 1957: *Treatise on Invertebrate Palaeontology. Part L. Mollusca 4. Cephalopoda Ammonoidea*. 490 pp. Geological Society of America and University of Kansas Press, Lawrence, Kansas.

Morton, S.G. 1829: Description of two species of fossil shells of the genera *Scaphites* and *Crepidula*; with some observations on the ferruginous sand formation, plastic clay and upper marine formations of the United States. *Journal of the Academy of Natural Sciences of Philadelphia 6*, 107–119.

Müller, G. & Wollemann, A. 1906: Die Molluskenfauna des Untersenon von Braunschweig und Ilsede. II. Die Cephalopoden. *Abhandlungen der Königlichen Preussischen Geologischen Landesanstalt, N.F. 47*, 30 pp.

Naidin, D.P. 1974: Ammonoidea. *In* Krymgolts, G.Ya. (ed.): *Atlas verkhnemelevoj fauny Donbassa [Atlas of the Upper Cretaceous Fauna of Donbass]*, pp. 158–195. Nedra, Moscow. [In Russian.]

Naidin, D.P. & Shimanskij, V.N. 1959: Cephalopoda. *In* Moskvin, M.M. (ed.): *Atlas verkhnemelovoj fauny severnogo Kavkaza i Kryma [Atlas of the Upper Cretaceous fauna of the northern Caucasus and Crimea]*, 166–220. Moscow. [In Russian.]

Neumayr, M. 1875: Die Ammoniten der Kreide und die systematik der Ammonitiden. *Zeitschrift der Deutschen Geologischen Gesellschaft 27*, 854–942.

Nilsson, S. 1826: Om de mångrummiga snäckor som förekomma i kritformationen i Sverige. *Kongliga Vetenskaps-Akademiens Handlingar 1825*, 329–343.

Nilsson, S. 1827: *Petrificata Suecana: Formationis Cretaceae*. 39 pp. Berling, Londini Gothorum.

Nowak, J. 1911: Untersuchungen über die Cephalopoden der oberen Kreide in Polen. II. Teil. Die Skaphiten. *Bulletin international de l'Academie des Sciences de Cracovie. Classe des Sciences Mathématiques et Naturelles. Série B. Sciences Naturelles*, 547–589.

Nowak, J. 1913: Untersuchungen über die Cephalopoden der oberen Kreide in Polen. III Teil. *Bulletin de l'Academie des Sciences de Cracovie. Classe des Sciences Mathématiques et Naturelles. Série B. Sciences Naturelles*, 335–415.

Nowak, J. 1916: Zur Bedeutung von *Scaphites* für die Gliederung der Oberkreide. *Verhandlungen der Geologischen Reichsanstalt (Staatsanstalt-Landesanstalt) for 1916*, 55–67.

Ødum, H. 1953: De geologiska resultaten från borrningarna vid Höllviken. *Sveriges Geologiska Undersökning C527*. 37 pp.

Orbigny, A. d' 1840–1842: *Paléontologie française: Terrains crétacés 1. Céphalopodes*. 1–120 (1840); 121–430 (1841); 431–662 (1842). Masson, Paris.

Orbigny, A. d' 1850: *Prodrome de Paléontologie stratigraphique universelle des animaux mollusques et rayonnés 2*. 428 pp. Masson, Paris.

Owen, D.D. 1852: *Report of a Geological Survey of Wisconsin, Iowa and Minnesota, and incidentally of a portion of Nebraska Territory: made under the direction of the US Treasury Department, Philadelphia*. 195 pp. Lippincott, Grambo & Co., Philadelphia, Pa.

Parkinson, J. 1811: *Organic Remains of a Former World, 3*. 479 pp. Robson, London.

Paulcke, W. 1907: Die Cephalopoden der oberen Kreide Südpatagoniens. *Bericht der Naturforschenden Gesellschaft zu Freiburg i. Breisgau 15, for 1905*, 167–248.

Ravn, J.P.J. 1902–1903: Molluskerne i Danmarks Kritaflejringer. *Kongelige Danske Videnskabernes Selskabs Skrifter (6) 11. 1, Lamellibranchiater*, 73–138 [5–70] (1902); 2, Scaphopoder, Gastropoder og Cephalopoder, 209–269 [5–65] (1902); 3, Stratigrafiske Undersøgelser, 339–433 [5–99] (1903).

Regnéll, G. 1983: Zoologen och arkeologen som var geolog. *In* Régnell, G. (ed.): Sven Nilsson. En lärd i 1800-talets Lund. *Studier utgivna av Kungliga Fysiografiska Sällskapet i Lund, 1983*, 23–83.

Reeside, J.B. Jr. 1927: Cephalopods from the lower part of the Cody Shale of Oregon Basin, Wyoming. *United States Geological Survey Professional Paper 150-A*, 1–19.

Riccardi, A.C. 1983: Scaphitids from the Upper Campanian – Lower Maastrichtian Bearpaw Formation of the western interior of Canada. *Geological Survey of Canada Bulletin 354*. 103 pp.

Riedel, L. 1931: Zur Stratigraphie und Faciesbildung im Oberemscher und Untersenon am Südrande des Beckens von Münster. *Jahrbuch der Preussischen Geologischen Landesanstalt zu Berlin 51*, 605–713.

Roemer, A. 1840–1841: *Die Versteinerungen des norddeutschen Kreidegebirges*. 145 pp. (1–48, 1840; 49–145, 1841). Hahn'schen, Hannover.

Roemer, F.A. 1865: Die Quadraten-Kreide des Sudmerberges bei Goslar. *Palaeontographica 13:4*, 193–199.

Rollier, L. 1922: Phylogénie des Ammonoïdes. *Eclogae Geologicae Helvetiae 17*, 358–360.

Roman, F. 1938: *Les ammonites jurassiques et crétacées. Essai de genera*. 554 pp. Masson, Paris.

Schlüter, C. 1867: *Beitrag zur Kenntniss der jüngsten Ammoneen Norddeutschlands*. 36 pp. Henry, Bonn.

Schlüter, C. 1870: Bericht über eine geognostisch-paläontologische Reise im südlichen Schweden. *Neues Jahrbuch für Mineralogie, Geologie und Paläontologie, Jahrgang 1870*, 929–969.

Schlüter, C. 1871–1876: Cephalopoden der oberen deutschen Kreide. *Palaeontographica 21*, 1–24 (1871); *21*, 25–120 (1872); *24*, 123–204, 207–263 36–55 (1876).

Schlüter, C. 1899: *Podocrates im Senon von Braunschweig und Verbreitung und Benennung der Gattung. Zeitschrift der Deutschen Geologischen Gesellschaft 51*, 409–430.

Schmid, F. & Ernst, G. 1975: Ammoniten aus dem Campan der Lehrter Westmulde und ihre stratigraphische Bedeutung. I. Teil. *Scaphites, Bostrychoceras* und *Hoplitoplacenticeras. Bericht der Naturhistorischen Gesellschaft zu Hannover 119*, 315–359.

Schulz, M.-G. 1878: Zur Litho- und Biostratigraphie des Obercampan–Untermaastricht von Lägerdorf und Kronsmoor (SW-Holstein). *Newsletters in Stratigraphy 7*, 73–89.

Schulz, M.-G. 1979: Morphometrisch-variationsstatistische Untersuchungen zur Phylogenie der Belemniten-Gattung *Belemnella* im Untermaastricht NW-Europas. *Geologisches Jahrbuch A47*, 157 pp.

Schulz, M.-G. 1996: Macrofossil biostratigraphy. *In* Schönfeld, J. & Schulz, M.-G. (Coord.) *et al.*: New results on biostratigraphy, palaeomagnetism, geochemistry and correlation from the standard section for the Upper Cretaceous White Chalk of northern Germany (Lägerdorf-Kronsmoor-Hemmoor). *Mitteilungen aus dem Geologisch-Paläontologischen Institut der Universität Hamburg 77*, 548–550.

Schulz, M.-G., Ernst, G., Ernst, H. & Schmid, F. 1984: Coniacian to Maastrichtian stage boundaries in the standard section for the Upper Cretaceous white chalk of NW Germany (Lägerdorf–Kronsmoor–Hemmoor): Definitions and proposals. *Bulletin of the Geological Society of Denmark 33*, 203–215.

Seunes, J. 1890–1892: Contributions à l'étude des céphalopodes du Crétacé Supérieur de France. 1. Ammonites du Calcaire à Baculites du Cotentin. *Mémoires de la Société Géologique de France, Paléontologie 1, Mémoire 2*, 1–7; *2, Mémoire 2*, 2–22.

Sharpe, D. 1853–57: Description of the fossil remains of Mollusca found in the Chalk of England. I. Cephalopoda. *Palaeontographical Society Monographs 68*, 1–26 (1853); 27–36 (1855); 37–68 (1857).

Sherborn, C.D. 1889: On the dates of the 'Paléontologie Française' of d'Orbigny. *Geological Magazine (iv) 6*, 223–225.

Shimizu, S. 1934: [*Ammonites.*] *In* Shimizu, S. & Obata, T.: [Cephalopoda. Iwanami's lecture series of Geology and Palaeontology.] 137 pp. [In Japanese.]

Siverson, M. 1992: Biology, dental morphology and taxonomy of lamniform sharks from the Campanian of the Kristianstad Basin, Sweden. *Palaeontology 35*, 519–554.

Smith, A.S. 1987: *Fossils of the Chalk*. 306 pp. Palaeontological Association, London.

Sowerby, J. 1812–1822: *The Mineral Conchology of Great Britain*. 383 pls. London.

Spath, L.F. 1922: On the Senonian ammonite fauna of Pondoland. *Transactions of the Royal Society of South Africa 10*, 113–147.

Spath, L.F. 1925: On Senonian Ammonoidea from Jamaica. *Geological Magazine 62*, 28–32.

Spath, L.F. 1926: On new ammonites from the English Chalk. *Geological Magazine 63*, 77–83.

Spath, L.F. 1927: Revision of the Jurassic Cephalopod fauna of Kachh (Cutch). *Memoirs of the Geological Survey of India, Palaeontologia Indica, N.S. 11*, 1–71.

Spath, L.F. 1953: The Upper Cretaceous cephalopod fauna of Graham Land. *Scientific Reports, Falkland Islands Dependencies Survey 3*, 1–60.

Stephenson, L.W. 1941: The larger invertebrate fossils of the Navarro Group of Texas (exclusive of corals and crustaceans and exclusive of the fauna of the Escondido Formation). *University of Texas Publication 4101*. 641 pp.

Stobaeus, K. 1732: *Opuscula in quibus petrefactorum, numismatum et antiquitatum historia illustratur, in unum volumen collecta*. 327 pp. Dantiscum.

Stolley, E. 1896: Einige Bemerkungen über die obere Kreide insbesondere von Lüneburg und Lägerdorf. *Archiv für Anthropologie und Geologie Schleswig-Holsteins 1*, 139–176.

Stolley, E. 1897: Ueber die Gliederung des norddeutschen und baltischen Senon sowie die dasselbe characterisierenden Belemniten. *Archiv für Anthropologie und Geologie Schleswig–Holsteins 2*, 216–302.

Stolley, E. 1930: Einige Bemerkungen über die Kreide Südskandinaviens. *Geologiska Föreningens i Stockholm Förhandlingar 52*, 157–190.

Stolley, E. 1932: Die Belemniten De Geer's von Båstad. *Geologiska Föreningens i Stockholm Förhandlingar 54*, 498–499.

Surlyk, F. 1984: The Maastrichtian Stage in NW Europe and its brachiopod zonation. *Bulletin of the Geological Society of Denmark 33*, 217–233.

Tate, R. 1865: On the correlation of the Cretaceous formations of the northeast of Ireland. *Quarterly Journal of the Geological Society of London 21*, 15–44.

Tomlin, J.R.B. 1930: Some preoccupied generic names. II. *Proceedings of the Malacological Society 19*, 22–24.

Wahlenberg, G. 1818: *Petrificata telluris Svecanae*. 116 pp. Uppsala.

Ward, P.D. & Kennedy, W.J. 1993: Maastrichtian ammonites from the Biscay Region (France, Spain). *Paleontological Society Memoir 34*. 58 pp.

Wegner, T. 1905: Die Granulatenkreide des westlichen Münsterlandes. *Zeitschrift der Deutschen Geologischen Gesellschaft 57*, 112–232.

Wiedmann, J. 1966: Stammesgeschichte und System der posttriadischen Ammonoideen; ein Überblick. *Neues Jahrbuch für Geologie und Paläontologie, Abhandlungen 125*, 49–79; *127*, 13–81.

Wollemann, A. 1902: Die Fauna der Luneburger Kreide. *Abhandlungen der Preussischen geologischen Landesanstalt N.F. 37*. 111 pp.

Woods, H. 1906: The Cretaceous fauna of Pondoland. *Annals of the South African Museum 4*, 275–350.

Worm, O. 1655: *Museum Wormianum*. 389 pp. Elsevier, Amsterdam.

Wright, C.W. 1952: A classification of the Cretaceous Ammonites. *Journal of Paleontology 26*, 213–222.

Wright, C.W. 1957: [Cretaceous Ammonoidea.] *In* Moore, R.C. (ed.): *Treatise on Invertebrate Paleontology. Part L, Mollusca 4, Cephalopoda Ammonoidea*. 490 pp. Geological Society of America and University of Kansas Press, Lawrence, Kansas.

Wright, C.W. 1979: The ammonites of the English Chalk Rock (Upper Turonian). *Bulletin of the British Museum (Natural History), Geology Series 31*, 281–332.

Wright, C.W. & Matsumoto, T. 1954: Some doubtful Cretaceous ammonite genera from Japan and Saghalien. *Memoirs of the Faculty of Science, Kyushu University, Series D, Geology 4*, 107–134.

Wright, C.W. & Wright, E.V. 1951: A survey of the fossil Cephalopoda of the Chalk of Great Britain. *Palaeontographical Society Monographs*, 1–40.

Zittel, K.A. von 1884: *Handbuch der Palaeontologie*. 1. Abt. 2; Lieferung 3, Cephalopoda, pp. 329–522. R. Oldenbourg, Munich & Leipzig.

Zittel, K.A. von 1895: *Grundzüge der Palaeontologie (Palaeozoologie)*. 972 pp. R. Oldenbourg, Munich & Leipzig.

9 788200 376958